Archaeological Pitchstone in Northern Britain

Characterization and interpretation of an important prehistoric source

Torben Bjarke Ballin

BAR British Series 476
2009

Published in 2016 by
BAR Publishing, Oxford

BAR British Series 476

Archaeological Pitchstone in Northern Britain

ISBN 978 1 4073 0386 4

BAR Publishing is the trading name of British Archaeological Reports (Oxford) Ltd.
British Archaeological Reports was first incorporated in 1974 to publish the BAR
Series, International and British. In 1992 Hadrian Books Ltd became part of the BAR
group. This volume was originally published by Archaeopress in conjunction with
British Archaeological Reports (Oxford) Ltd / Hadrian Books Ltd, the Series principal
publisher, in 2009. This present volume is published by BAR Publishing, 2016.

Printed in England

BAR
PUBLISHING

BAR titles are available from:

BAR Publishing
122 Banbury Rd, Oxford, OX2 7BP, UK
EMAIL   info@barpublishing.com
PHONE   +44 (0)1865 310431
FAX   +44 (0)1865 316916
www.barpublishing.com

TABLE OF CONTENTS

LIST OF COLOUR PLATES

LIST OF FIGURES

LIST OF TABLES

ACKNOWLEDGEMENTS

Over the almost four years it took to bring the Scottish Archaeological Pitchstone Project to its conclusion, I received funding and practical help or advice from most quarters of Scottish archaeology as well as from institutions and individuals in the wider UK. Whether people or institutions offered their support because this was part of their remit, or because they found the subject matter as fascinating as I do, I am equally grateful to everybody for their help. Scottish archaeological pitchstone and its distribution – not least the interpretation of it – is a complex matter, and I would not have been able to carry out this project without the support people generously offered.

The Scottish Archaeological Pitchstone Project received funding from three of the main Scottish institutions, namely Historic Scotland, National Museums Scotland, and the Society of Antiquaries of Scotland. The grants provided by these three bodies made it possible to undertake the initial planning, visit pitchstone-holding museums, produce the pitchstone database and, finally, write up and publish the project. During this process, I received continued encouragement from Principal Inspectors Patrick Ashmore and Noel Fojut from Historic Scotland, and Senior Curator Alan Saville from National Museums Scotland. The Robert Kiln Trust generously provided the means to allow the examination and publication of the pitchstone from Biggar in South Lanarkshire, and the Catherine Mackichan Bursary Trust made it possible to visit the Glencallum Bay pitchstone source on Bute in connection with the investigation and publication of the porphyritic pitchstone assemblage from Blackpark Plantation East on that island.

As described in Chapter 1, the project resulted in the publication of a number of 'spin-off' papers. Some of these were produced in cooperation with other 'pitchstone enthusiasts', and without their input these papers would have been much less informative. Geologist John Faithfull from the Hunterian Museum and Art Gallery, Glasgow, was co-author of the Arran pitchstone gazetteer and the Blackpark Plantation East paper, and his knowledge on geological pitchstone has been immensely valuable to the production of these two papers and to the project in general. Tam Ward, who is the Curator of Biggar Museum and the leader of the Biggar Museum Archaeology Group, co-authored the Biggar pitchstone paper, and he has personally been involved in the recovery of most of the pitchstone from South Lanarkshire. Archaeologist Chris Barrowman initially discovered the Blackpark Plantation East site during fieldwalking, and he recovered a large proportion of the presently known material from that site. He also took part in the publication of the assemblage, with the author of the present volume and John Faithfull.

Along the way, I have been in regular contact with a number of colleagues, who kindly offered general advice within their areas of expertise. Senior Curator Alan Saville from National Museums Scotland has always been available for questions regarding lithic issues, and he has commented on sections of this report; Alison Sheridan, Head of Early Prehistory, National Museums Scotland, has been equally generous with her time in connection with questions regarding the Early Neolithic period in general, as well as its various pottery styles; and John Faithfull, the Hunterian Museum and Art Gallery, Glasgow, set aside much time for the discussion of geological pitchstone, on and off Arran.

During the project, numerous – if not most – Scottish museums and stores were visited or contacted, and the following is a list of pitchstone-holding institutions which have been helpful in connection with the compilation of the pitchstone database (the institution's contact person is in brackets): Historic Scotland (Jane Flint); National Museums Scotland (Alan Saville); the Royal Commission on the Ancient and Historical Monuments of Scotland (Steve Boyle); the Hunterian Museum and Art Gallery (John Faithfull, Sally-Anne Coupar); the Kelvingrove Art Gallery and Museum (Katinka Stentoft); Biggar Museum (Tam Ward); Dumfries Museum (Joanne Turner); Bute Museum (Anne Speirs); Stranraer Museum (John Pickin); Hawick Museum (Richard White); Marischal Museum (Neil Curtis); the MacLean Museum, Greenock (Val Boa); Paisley Museum (Donald Macleod); Arran Museum (Margaret Bruce); Orkney Museums (Anne Brundle); Perth Museum (Mark Hall); Kilmartin House Museum (Sharon Webb); Aberdeenshire Heritage (David Bertie); Fife Museum Service (Niamh Conlon); Tain and District Museum (Alistair Jupp); Stirling Smith Art Gallery and Museum (Michael McGinnes); Montrose Museum (Linda Fraser); and Cambridge University Museum of Archaeology and Anthropology (Anne Taylor).

To cover the latest pitchstone finds from regions with few or no finds of archaeological pitchstone, it was necessary to visit or contact a number of regional archaeological services, as well as archaeological units. I am indebted to the following for their invaluable contributions: Orkney Archaeological Trust (Jane Downes, Nick Card, Julie Gibson); Glasgow University Archaeological Research Division (Aileen Maule, John

Atkinson, Beverley Ballin Smith); CFA Archaeology Ltd. (Sue Anderson); Headland Archaeology Ltd. (Julie Franklin); AOC Archaeology Group (Ciara Clarke, Rob Engl); SUAT Ltd. (Catherine Smith); and Murray Archaeological Services Ltd. (Hilary and Charlie Murray).

In many cases, it was necessary to contact archaeologists, geologists or archaeological enthusiasts to enquire further into specific issues, assemblages or locations, and I am grateful to the following for taking the time to deal with my questions or lending me assemblages or individual finds: Kristian L.R. Pedersen (access to the latest Northumberland pitchstone); Lawrence Abbott (off-prints of relevance to ethnographic issues); the Implement Petrology Group (helping to organize the search for English pitchstone); Alastair Buckoke (access to the assemblage from the Kilhern Bog Site); Diana Coles (access to the assemblage from the Knocknab site); Euan MacKie (information on the finds from the Duntreath Standing Stones); Andrew Hendry (access to the finds from Cairnpapple); Rod McCullough (information on the pitchstone from Newton, Islay); Vicki Cummings (information on the latest pitchstone finds from Kintyre); Richard Bradley (access to his finds from Ben Lawers); Douglas Baird (information on the pitchstone from Little Gight); Clive Bonsall (information on the pitchstone from Ulva Cave); Caroline Wickham-Jones (information on the pitchstone from Camas Daraich); Graeme Warren (information on the pitchstone from Warrenfield and Monkton); Mike McCarthy (information on the pitchstone from Carlisle); Pete Topping (advice on ethnographical matters); Anastasia Steffen (advice on obsidian and permission to use illustrations from her dissertation); Derek Roe (permission to use Figure 16); Elsa Henderson (thin-sections of Bute pitchstone); Martin Lee (thin-sections of Bute pitchstone); Andy Jones (information on the pitchstone from Torbhlaren); Trevor Cowie (information on the pitchstone from Carwinning Hill); Derek Alexander (information on the pitchstone from Deer's Den); Rob Engl (information on the pitchstone from Cambuslang); Martin Cook (information on the pitchstone from Nether Largie); Jeremy Preston (various comments on geological pitchstone); Jonathan Wordsworth (information on the pitchstone from Upper Cullernie); Alex Morrison (information on the pitchstone from Houston South Mound); and Tom Clare and Peter Cherry (information on potential pitchstone from Cumbria).

Last, but definitely not least, I would like to thank my wife, Beverley Ballin Smith, for constant encouragement through the project, as well as for practical help in connection with the photographing of artefacts and for final checks of my spelling, grammar and syntax.

1. Introduction

1.1 Background

In 1984, Williams Thorpe & Thorpe published their now widely cited paper on the distribution and sources of archaeological pitchstone in northern Britain. Based on chemical analysis of archaeological pitchstone samples, and comparison with similarly analysed samples of geological pitchstone from the Palaeogene (formerly Tertiary) Volcanic Districts of Scotland (cf. Emeleus & Bell 2005; Richey 1961), it was concluded that most, if not all, archaeological pitchstone derives from the island of Arran in the Firth of Clyde (Figure 1). The paper included an appendix in which all archaeological sites with pitchstone were listed, and thoughts were put forward regarding the socio-economic mechanisms behind the observed distribution pattern.

Now, a quarter of a century later, many more pitchstone artefacts have been recovered. When Williams Thorpe & Thorpe's paper was published, only approximately 1,400 pitchstone artefacts from *c*. 100 find locations were known (Williams Thorpe & Thorpe 1984, Fig. 2), and no more than four sites were mapped north of the Firth of Tay (one in the Highland region and three in the northern part of the Grampian region). The majority of the remaining archaeological pitchstone derived from Arran itself, the Tweed valley or the area around Luce Bay in Dumfries & Galloway.

Today (2008), the number of pitchstone-bearing sites has multiplied several times and approximately 20,300 worked pieces from *c*. 350 sites have been found (see Appendices 1 and 2), and pitchstone artefacts have been reported from practically all parts of Scotland (apart from Shetland), as well as from northern England (Burgess 1972; Fell 1990; Kristian L.R. Pedersen pers. comm.), Northern Ireland (Simpson & Meighan 1999), and the Isle of Man (McCartan & Johnson 1991). Most of the new locations represent excavated material with well-defined find contexts, whereas Williams Thorpe & Thorpe's list included many stray finds with low research potential.

1.2 Aims and Objectives

As a response to this extraordinary increase in the volume of evidence, a new project was defined - the Scottish Archaeological Pitchstone Project - which in the following text is abbreviated to 'SAPP'. Its main aims were:

- To update Williams Thorpe & Thorpe's list of archaeological pitchstone in northern Britain (1984) by producing an Access database of all presently (2008) known pieces of archaeological pitchstone and, on the basis of the now much larger volume of evidence;
- To clearly characterize pitchstone as a raw material or 'tool stone' (geology/provenance, procurement, flaking properties, operational schemas, etc.);
- To discuss the dating of the archaeological evidence;
- To re-interpret the dispersal of archaeological pitchstone across northern Britain, as well as the distribution's socio-economic background, by defining a territorial structure and exchange network for this region.

CD copies of the database will be lodged with the National Monuments Record of Scotland in Edinburgh, National Museums Scotland and other relevant museums in the region.

The present project offers several benefits to Scottish museums:

- As most of the archaeological pitchstone is in the care of a small number of museums (National Museums Scotland, the Kelvingrove Art Gallery and Museum, the Hunterian Museum and Art Gallery, Biggar Museum, and Dumfries Museum), the computer database will represent an authoritative catalogue of the pitchstone holdings of these institutions;
- The project sets a standard in characterizing pitchstone artefacts that will allow museums to characterize and classify their pitchstone artefacts more effectively in the future.

1.3 Working Hypotheses

From very early in the project, the growing volume of evidence allowed the formulation of a number of broad working hypotheses, based on 'impressions' of the examined archaeological material. At the most general level, it was possible to conclude 1) that much of what had become general consensus in the field of archaeological pitchstone research had an almost mythical character, and that very little of our 'knowledge' reflected reality accurately, and 2) that archaeological pitchstone was distributed across northern Britain in a non-random fashion.

1. North Ayrshire
2. Inverclyde
3. Renfrewshire
4. West Dunbartonshire
5. East Dunbartonshire
6. Glasgow
7. East Renfrewshire
8. North Lanarkshire
9. Falkirk
10. Clackmannanshire
11. West Lothian
12. Edinburgh

A. Luce Bay, Dumfries & Galloway
B. Biggar, South Lanarkshire
C. Blackpark Plantation East, Bute
D. Ballygalley, Antrim,

Figure 1. Location map, showing the countries and counties of northern Britain. The smaller counties of the Scottish Central Belt are labelled 1-12, and the four largest concentrations of archaeological pitchstone are labelled A-D.

When the SAPP was initiated in 2005, 'general consensus' included simplistic claims such as: 1) 'all pitchstone artefacts off Arran are based on aphyric material from outcrops in the Corriegills district'; 2) 'Scottish pitchstone artefacts generally date to the Mesolithic, Neolithic and Bronze Age periods'; 3) 'there are practically no cores and tools in pitchstone assemblages off Arran, and in these areas most finds represent unmodified waste', and 4) 'except on Arran and perhaps in the Luce Bay area (Dumfries & Galloway), pitchstone played a minor role in stone tool assemblages' (eg, Ritchie 1968; Williams Thorpe & Thorpe 1984). Following the production of the database, it has been possible to partially reject or adjust all of these claims (see the following chapters), and more accurately address questions such as:

- Pitchstone chronology, on and outwith Arran;
- General raw material preference on and outwith Arran (aphyric/porphyritic pitchstone);
- The composition of pitchstone assemblages;
- Pitchstone reduction methods, for example whether pitchstone was reduced in the same way as related materials, or whether pitchstone knapping required specialized approaches;
- Whether pitchstone was reduced on the mainland or imported from Arran in the form of final blanks and tools.

Summary examination and analysis of *c.* 1700 pieces of worked pitchstone in the stores of National Museums Scotland (SAPP Phase 1: 2006) indicated a zonation of northern Britain, with the most extensive exploitation of pitchstone occurring on the source island, Arran, and with a number of zones surrounding it (Chapter 4.2). These zones are defined by a combination of decreasing assemblage size and decreasing frequency of pitchstone-bearing sites with increasing distance to the raw material outcrops on the Isle of Arran. One of the key purposes of the SAPP was to test the existence of these zones and define their number and character as part of the general discussion of the distribution of archaeological pitchstone.

In Chapter 7.5, the discussion focuses on whether the division of northern Britain represents the rudiments of prehistoric territorial structures: where the division of Scotland into regions based on the exploitation of quartz and flint/flint-like materials may represent different techno-complexes (Ballin 2005c), the pitchstone zones most likely represent different social territories, that is, territories with varying access to the pitchstone sources on Arran and with different perceptions (functional/symbolic) of pitchstone.

1.4 Project History

The SAPP was formally structured as a sequence of phases and, parallel to the tasks defined for these phases, a number of special case studies were carried out. In total, the project

included four main phases, initially defined in the following manner:

- Phase 1 (Jan – March 2006): the examination and characterization/cataloguing of all pitchstone held by National Museums Scotland;
- Phase 2 (Jan – March 2007): the examination and characterization/cataloguing of all pitchstone held by the Hunterian Museum and Art Gallery and the Kelvingrove Art Gallery and Museum;
- Phase 3 (Oct 2007 – Jan 2008): the examination and characterization/cataloguing of all pitchstone held by smaller, local Scottish museums;
- Phase 4 (April 2008 – March 2009): production of the project report (the present volume).

In practice, tasks were moved between phases as time permitted or circumstances made necessary. During Phase 1, not only the holdings of National Museums Scotland were examined, but also that of the Orkney Museum (the finds from Barnhouse; the finds from Ness of Brodgar were examined during Phase 3). During Phase 2, the holdings of the main Glasgow museums were examined, and also those of the Marischal Museum in Aberdeen, Dumfries Museum, Biggar Museum, and Bute Museum. In the course of Phase 3, the remaining smaller museum collections were examined (Paisley Museum, the McLean Museum, Hawick Museum, Stranraer Museum; National Museums Scotland and the Kelvingrove Art Gallery and Museum were revisited), but it was also deemed necessary to move beyond museum collections, and a number of commercial archaeological units, individual archaeological analysts, and local enthusiasts were contacted.

Initially, it was deemed too time-consuming to contact the excavating units, individual researchers or local enthusiasts/ societies, but at a later stage this approach was thought to be essential to cover special cases, simply to make the database truly geographically representative, and to increase its research potential. A good example of such a situation is the distribution of pitchstone throughout Ayrshire, where there is a noticeable discrepancy between the number of pitchstone artefacts known after the examination of the final museum collections (approximately 20), and the number known after the examination of assemblages in the temporary care of GUARD and individual, mostly university-based, researchers (134 pieces). This case probably reflects the generally low level of development (housing, roads, motorways, railways, etc.) in Ayrshire pre *c.* 1990 compared to that in the rest of Scotland, as well as a perceptible increase of development projects in Ayrshire in more recent years. It demonstrates how geographically/ topographically unevenly distributed research may skew the interpretation of prehistoric societies (Chapter 7.2). Another example is the recent, and as yet unpublished, recovery of worked pitchstone on Ben Lawers in Perth & Kinross, by colleagues from GUARD, the RCAHMS, and the University of Reading (Richard Bradley), which shows how pitchstone penetrated Scotland and reached what would traditionally be perceived as peripheral areas of the country.

During Phase 4, the present manuscript was produced. As a large number of approaches to smaller museums, units, colleagues, and amateurs remained unanswered at the end of Phase 3, this phase was also used to update, amend, supplement, and test-run/'polish' the database. As a consequence of a number of published papers and notes presenting the results of the SAPP (Ballin 2006b; 2007a; 2008a; 2008e), new finds of archaeological pitchstone continued to be reported to the author and these finds were added to both the database and the distribution maps used in the discussion of the dispersal of archaeological pitchstone (Chapter 7.5).

During the course of Phases 1-4, a number of particularly interesting individual assemblages were encountered, and it was decided to analyse these in greater detail and publish them separately. It was also deemed necessary to look into the general question of pitchstone provenance, as it was the author's impression that the claimed exclusivity of the Corriegills outcrops represented an exaggeration, not least in connection with the interpretation of the situation on Arran itself. Consequently, the following sub-projects were carried out in parallel with Phases 1-4, and funded separately (see the project's acknowledgements):

- The assemblage from *Auchategan, Glendaruel, Argyll & Bute* (comparison between the operational schemas of the pitchstone, flint and quartz sub-assemblages; Ballin 2006a; also see Marshall 1978);
- The assemblage of *Barnhouse, Orkney* (the dating, characterization and interpretation of an unusually large, 'peripheral' pitchstone assemblage; Ballin forthcoming b);

- The assemblages of *the Biggar Area, South Lanarkshire* (the characterization and interpretation of a potential pitchstone 'redistribution' centre; Ballin & Ward 2008);
- The assemblage of *Blackpark Plantation East, Bute* (the dating, characterization and interpretation of an unusually large assemblage of heavily porphyritic pitchstone; Ballin *et al.* forthcoming);
- *Gazetteer of Arran pitchstone outcrops* (termination of the myth that Tyrrell's (1928, 229) four main pitchstone outcrops – Corriegills, Glen Shurig, Glen Cloy and Tormore – are the only archaeologically relevant sources on Arran; Ballin & Faithfull forthcoming).

Over the approximately four year duration of the SAPP, it was also possible to analyse smaller pitchstone-bearing assemblages as they became known to the author. These assemblages include:

- From CFA Archaeology Ltd.: Auchrannie (Arran), East Lochside (Angus), and Dalkeith Northern Bypass (Midlothian);
- From GUARD: Midross, Loch Lomond (Argyll & Bute), Blackshouse Burn (South Lanarkshire), Urquhart Castle (Highland), and Laigh Newton (East Ayrshire);
- From Headland Archaeology Ltd.: St Marnock's Chapel, Inchmarnock (Argyll & Bute);
- From SUAT Ltd.: Achnahaird Sands (Highland);
- From St Andrews Heritage Services: Fordhouse Barrow (Angus);
- From National Museums Scotland: Meldon Bridge (Scottish Borders), Lussa River, and Lealt Bay (the latter assemblages are both from Jura, and this research was carried out in connection with the author's Quartz Project; Ballin forthcoming j).

2. ARCHAEOLOGICAL VOLCANIC GLASS
– A BRIEF RESEARCH HISTORY

2.1 Scottish Pitchstone Research

2.1.1 Introduction

The first references to Scottish pitchstone appear in the literature of the late eighteenth century. Although the early focus may mainly have been geological (Jameson 1798), it was noted (Robertson 1768, in Mitchell 1898, 13, 18) that, near Kilbride on Arran, '... there is an uncommon kind of rock with which the ancient inhabitants tipped their arrows'. Over the following century, the interest in Scottish pitchstone grew steadily, but it was still largely confined to the field of geology. The period from the mid 1800s to approximately 1930 saw the publication of most of the major geological works on pitchstone (eg, Bryce 1859; Judd 1893; Gunn *et al.* 1903; Tyrrell 1928). After 1930, the interest in geological pitchstone weakened considerably, only to be re-kindled fifty years later. Much of modern-day geological pitchstone research concerns the development of methods for provenancing pitchstone, and a proportion of this work has been carried out in an archaeological context (eg, Williams Thorpe & Thorpe 1984; Preston *et al.* 1998; Preston *et al.* 2002). For an overview of geological research into Scottish pitchstone, and a discussion of pitchstone as a geological material, see Ballin & Faithfull (forthcoming).

Archaeological research into pitchstone picked up at the beginning of the twentieth century (eg, Mann 1918) and the work immediately concentrated on two main areas of interest, namely 1) pitchstone recovered from archaeological sites on Arran, and 2) the distribution of artefacts in Arran pitchstone across the adjacent parts of mainland Scotland (with the publication of first Ritchie 1968, and later Williams Thorpe & Thorpe 1984, this geographical area was expanded to northern Britain in general).

2.1.2 Pitchstone from Sites on Arran

The investigation of archaeological pitchstone on Arran commenced with Bryce's study of the island's megalithic cairns (1902; 1903; 1909; 1910). Bryce noticed that many assemblages from Arran cairns included pitchstone artefacts, and it was an obvious thought that these pieces were burial goods in the traditional sense. However, later excavation of Arran cairns, and not least their surroundings, presented a more complex picture (eg, MacKie 1964).

Although, on Arran, relatively sophisticated pitchstone artefacts may be recovered from megalithic chambers, unmodified 'waste' is occasionally found in entrance areas and forecourts, or underneath the burial monuments. The latter phenomenon is mirrored outside Arran, where simple pitchstone flakes or debris have been retrieved from, for example, pre-barrow pits (eg, Fordhouse Barrow in Angus; Ballin forthcoming f). It appears that some pitchstone may have been deposited as 'conventional' burial goods, whereas the deposition of pitchstone debris under or around burial monuments may have had other functions, possibly relating to symbolic values associated with the material's distinctive colour (very much in the same way as crushed white quartz was strewn across or around burial monuments in other parts of Scotland; Lebour 1914; Warren & Neighbour 2004; Ballin forthcoming j). Henshall's *The Chambered Tombs of Scotland Vol 2* (1972) provides a useful overview of Arran burial monuments with pitchstone finds, and finds were also recovered from burial monuments in connection with Barber's (1997) work on the island.

Although the Arran stone circles were noted and excavated by early antiquarians (Bryce 1862), and again through the twentieth century (eg, Roy *et al.* 1963), substantial numbers of pitchstone artefacts were not recovered from their surroundings until Haggarty's (1991) work at Stone Circles I and XI at Machrie Moor. However, most of these pieces appear to either pre- or post-date the activities at the stone circles and their timber-built predecessors, with many finds deriving from plough soil or postholes. It is uncertain what proportion of the lithic finds from the stone circles are primary deposits, and which are residual 'background noise'.

Domestic sites on Arran yielding pitchstone are well-known, partly as a result of early and later antiquarian research and partly as a result of recent developmental and forestry work on the island. Mann (1918, 144) mentioned a workshop at the Broddick Schoolhouse outcrop, but over the following decades focus concentrated on Arran's burial monuments rather than domestic sites (above). However, during the second half of the twentieth century this changed. Allen & Edwards (1987) presented a group of new individual finds and sites, many of which are clearly domestic. A large proportion of these locations were discovered in connection with forestry work, including the site of Auchareoch, which was later excavated and published by Affleck *et al.* (1988).

Barber's (1997) investigation of the Arran landscape was also instigated by forestry work, and a large number of domestic sites were discovered; the pitchstone was examined by Nyree Finlay. As mentioned above, much of the pitchstone from Haggarty's (1991) excavations at the Machrie Moor stone circles are thought to be residual domestic waste. Within the next few years, a number of large and small domestic sites yielding pitchstone are to be presented in connection with the publication of GUARD's work along the Arran Ring Main Water Pipeline (Donnelly & Finlay forthcoming). This block of mainly east Arran sites will be a fine supplement to the group of mainly west Arran sites produced by Barber.

2.1.3 Pitchstone from Sites beyond Arran

From early on, archaeologists were aware that Arran pitchstone had been exported to the Scottish mainland, either as raw tabular pieces (to be reduced at the destination) or as blanks and final tools. Smith (1897) mentioned the presence of pitchstone in Shewalton on the mainland opposite Arran, and Bryce (1903) included sites on Bute in his discussion of Arran megalithic tombs. Mann (1918) dedicated an entire chapter to the discussion of pitchstone 'trade', and he listed Bute, Ayrshire and Wigtownshire as counties with pitchstone-yielding sites. He also discussed the possibility of pitchstone being recovered from Ireland, but he concluded that the pieces of 'jet-black rock' he had seen are most likely dark chert.

Following Mann's paper, the distribution 'network' of archaeological pitchstone was slowly expanded. Lacaille (1931, 268) noticed a piece of worked, probably residual pitchstone in the assemblage from the Early Bronze Age cemetery at Cowdenbeath in Fife: '... the first recorded example from a locality north of the Forth'. Other pitchstone-bearing sites soon followed, with further examples from the south-west (eg, Lacaille 1945), and sites were added from the south-east (eg, Mulholland 1970), the east (see Table 2-3 in Warren 2006), the west (eg, Marshall 1978; Tolan-Smith 2001); the north (Corcoran 1966; Sharples 1981); and from the Western Isles (Warren 2003) and Orkney (Middleton 2005). Over the last couple of decades, exceptional amounts of worked pitchstone have been recovered by members of the Biggar Museum Trust archaeology group, including the discovery of several sites in South Lanarkshire with more than 100 pieces (see lists in Ness & Ward 2001; Ballin & Ward 2008). A small number of pitchstone-yielding sites are known from Northern Ireland, but recently more than 500 pieces were recovered from the site of Ballygalley in County Antrim (Simpson & Meighan 1999). A small number of assemblages with pitchstone are known from northern England (Fell 1990; Burgess 1972; Ritchie 1968; Kristian L.R. Pedersen pers. comm.), and pitchstone artefacts have also been reported from the Isle of Mann (McCartan & Johnson 1991).

In addition to Mann's 1918 paper, two synthetic papers

on pitchstone distribution and exchange were published (Ritchie 1968; Williams Thorpe & Thorpe 1984). All three papers include scientific provenancing of pitchstone and reached the conclusion that most worked pitchstone found beyond Arran is aphyric material, and therefore most likely to derive from the Corriegills district, but that a proportion of the pitchstone, such as the finds from Dunagoil on Bute (Mann 1918, 146), derive from porphyritic sources on Arran, such as the outcrops in Glen Shurig and Glen Cloy, and at Tormore.

The presently available material suggests that the vast majority of archaeological pitchstone found beyond Arran is from the Early Neolithic period (Chapter 5). It has not been possible to positively date any non-Arran pitchstone to the Mesolithic period, although the fact that many Early Neolithic pitchstone blades are fairly narrow (like for example the finds from Auchategan and Fordhouse Barrow; Marshall 1978; Ballin 2006a; forthcoming f) convinced many early antiquarians that a proportion of this material was manufactured by hunter-gatherers. One Late Neolithic chisel-shaped arrowhead has been found in the Glenluce Sands area of Dumfries & Galloway (registered as a triangular arrowhead in Williams Thorpe & Thorpe 1984, catalogue no. 48) and one in the Biggar area (Ballin & Ward 2008), but no secure find contexts or diagnostic types/attributes suggest the general use of pitchstone beyond Arran and Argyll & Bute (see Chapter 7.5) after the Neolithic period.

Like on Arran itself, mainland pitchstone artefacts have been recovered from burial and ritual sites as well as from domestic settlements. Pitchstone-yielding burial, ritual and domestic sites have been reported from all parts of mainland Scotland, and – apart from the varying number of pitchstone artefacts found at varying distances from Arran – it has not been possible, as yet, to define any regional differences in deposition practices. Henshall's *The Chambered Tombs of Scotland Vol 1 and Vol 2* (1963; 1972) provides an overview of non-Arran burial monuments with pitchstone finds.

2.2 Obsidian Research and its Relevance to the Study of Scottish Pitchstone

Internationally, the research of archaeological volcanic glass has always focused on obsidian (the distinction between obsidia and pitchstone is explained in Chapter 3), although the almost complete absence of references to pitchstone in the international archaeological literature suggests that, in some cases, the term 'obsidian' may cover obsidian proper as well as its 'cousin' pitchstone. It is almost impossible to distinguish between the purest aphyric pitchstones (such as some of the material from the Dun Fionn area in the 'greater' Corriegills district; Ballin & Faithfull forthcoming) and common obsidian, and obsidian and aphyric pitchstone probably flake in the same manner, and had the same use-value / symbolic connotations for prehistoric people.

Obsidian sources are known from all continents (see the IAOS Source Catalogue at http://www.peak.org~obsidian/index.html), and the research into obsidian has obviously been defined by local needs, and it followed local archaeological traditions. An overview of this work can be obtained by consulting the IAOS Obsidian Bibliography at the same Web address as the IAOS Source Catalogue (above). The questions dealt with in archaeological obsidian literature generally correspond to the questions dealt with in papers on assemblages in other raw materials, but a number of issues are typical of obsidian research, due to attributes particularly characteristic of this material. Those issues are raw material dating and provenancing.

Contrary to assemblages in most other lithic raw materials, which are usually dated by association with other datable materials or by context, it is possible to date obsidian directly. The method used for estimating the absolute date of obsidian artefacts is called obsidian hydration, which is based on the fact that obsidian, when fractured, starts absorbing water at a rate characteristic of specific obsidian sources (for details, see Renfrew & Bahn 1996, 150).

Contrary to for example flint, which it is still almost impossible to provenance securely (although see Högberg & Olausson 2008 for an account of recent Scandinavian research), a whole raft of scientific methods are available for the provenancing of obsidian, including optical emission spectrometry, neutron activation analysis, atomic absorption spectrometry, X-ray fluorescence spectrometry, particle-induced gamma-ray emission, and proton-induced gamma-ray emission, all of which are forms of trace element analysis; another approach is fission-track analysis, which is primarily a dating method, but it has also been used to distinguish between obsidian from different sources (for an overview of these approaches, see Renfrew & Bahn 1996, 346; Thatcher 2001, 48). For these reasons, many obsidian papers include extended discussions of assemblage date and regional aspects (such as, territoriality and exchange).

In terms of dating, it has yet to be tested whether hydration dating might be possible in connection with the investigation of Scottish pitchstone-yielding assemblages. However, the considerably higher water content of pitchstone (generally, up to 10 times as much as that found in common obsidians) may rule out the effectiveness of hydration dating of pitchstones (Jeremy Preston pers. comm.).

In terms of provenancing, the questions asked in connection with obsidian research and pitchstone research differ considerably. In for example Mediterranean obsidian research, the main task has been to assign raw material from obsidian-yielding sites to specific regions, for example by linking obsidian from Near Eastern sites to outcrops in Armenia or Anatolia (Renfrew *et al.* 1968), or by linking obsidian from Italian sites with outcrops on either Lipari or Sardinia (Acquafredda & Muntoni 2008). In Scottish pitchstone research, the region of origin is well-known, as various tests have shown that most, if not all, archaeological pitchstone from sites in northern Britain derives from the Isle of Arran, in the Firth of Clyde (Williams Thorpe & Thorpe 1984; Preston *et al.* 1998; Simpson & Meighan 1999). It may be possible to link pitchstone artefacts from the Scottish mainland to individual outcrops on Arran (eg, Ballin & Faithfull forthcoming), but this would move the research emphasis from the general northern British exchange network responsible for the dispersal of pitchstone (the present project) to the internal socio-economical organization of Arran itself (cf. Renfrew 1976).

To the SAPP, the most important methodological approach borrowed from obsidian research may be the use of regression analysis, or fall-off curves, in the analysis of exchange systems (Chapter 7). This approach has been applied to other types of products (see overview in Hodder 1974; Hodder & Orton 1976; Orton 1980; 120) but, within lithic studies, it has probably been most successfully used to shed light on Near Eastern obsidian exchange (Renfrew 1977; Renfrew *et al.* 1968).

3. PITCHSTONE TYPES

3.1 Geological Pitchstone

Scottish pitchstone is a member of the family of silica-rich volcanic glasses, and it is closely related to obsidian. The main difference between the two raw materials is the fact that pitchstone has a lustre like broken pitch (hence the name), whereas obsidian is distinctly vitreous. Pitchstone also contains considerably more water than obsidian. Based on Le Maitre (2002), Ballin & Faithfull (forthcoming) defined pitchstone as hydrated glassy rocks (typically 3-10% H_2O), while obsidians are nearly anhydrous (< 1% H_2O); most pitchstones have > 5% H_2O, and most obsidians < 0.5% H_2O.

Pitchstone may be characterized in terms of a number of components, the most important of which are: 1) a glassy matrix; 2) phenocrysts; 3) spherulites; and 4) crystallites. Phenocrysts are macroscopic, isolated or clustered crystals formed at depth during slow cooling; spherulites are finely crystalline, usually radiating intergrowths of quartz and feldspar indicating devitrification of the glass; and crystallites are very small skeletal or dendritic crystals, often Fe-Mg silicates, in the glass (banding in pitchstone is often marked by variation in crystallite density). Plates 1 to 3 show a spectrum of aphyric-porphyritic pitchstone.

Arran pitchstone often shows flow-banding, which is a consequence of the very high viscosity of the original melts, resulting in laminar flow during volcanic intrusion, and hence little mixing or homogenization compared with a turbulently convecting, low-viscosity magma. The banding is almost always emphasized by weathering, or other alteration (for example exposure to fire; Chapter 3.2.2). The colour of pitchstone is frequently referred to as dark-green, but as explained by Tyrrell (1928, 206), Arran pitchstone forms a continuum of colours '... from a light yellowish-green, through various dark shades of the same colour, to a black rock'. Most archaeological specimens are black with a green tinge.

Tyrrell (1928) suggested that it would be possible to identify most Arran pitchstones as belonging to one of four main categories, which he defined on the basis of textures and mineralogy: *Corriegills* (aphyric); *Glen Shurig* (quartz, feldspar, fayalite and pyroxene phenocrysts, with fayalite ≥ pyroxene), and *Tormore* (like Glen Shurig, but pyroxenes are much more abundant than fayalite). The chemistry of the *Glen Cloy* pitchstone type differs somewhat from the above, as it is poorer in silica, and much darker in colour, even under the microscope. However, there are many rocks which do not fit this scheme, such as pitchstones with only quartz and feldspar phenocrysts, or those where orthopyroxene is dominant, or some highly porphyritic types. As there is a continuum between many of the groups, and as the range of rocks is much greater than Tyrrell's classification suggests, Ballin & Faithfull (forthcoming) suggested that his scheme should be abandoned in favour of simple descriptive names based on texture, phenocryst assemblages and glass composition.

Pitchstone can occur geologically in a variety of environments. They result from the rapid cooling of silica-rich magmas (the same magmas as give rise to granitic rocks and rhyolites). Such rapid cooling is restricted to surface and near-surface geological settings. Pitchstones can therefore form as lavas or as shallow-level intrusions. Although the large Sgurr of Eigg pitchstone is a lava flow or, more likely, an ignimbrite (Brown *et al.* 2007), most Scottish occurrences, including all the Arran ones, are intrusive sheets. Some are sub-horizontal sheets or sills, while some are vertical dykes. Frequently, they occur as 'composite intrusions', usually with a pitchstone centre and margins of basalt or a similar rock. The pitchstones of Arran are widespread, but they are clearly a late feature of igneous activity, as they are found cutting most of the other Tertiary volcanic rocks. Various scientific methods for the characterization and provenancing of pitchstone are presented and discussed in Ballin & Faithfull (forthcoming).

3.2 Forms of Pitchstone 'Look-Alikes' and Altered Pitchstone

During fieldwork, Scottish archaeologists occasionally encounter artefacts in materials which either look like pitchstone but are not, or which are pitchstone but deviate visually from the well-known forms of volcanic glass. Below, these two groups are briefly characterized and discussed.

3.2.1 Pitchstone 'Look-Alikes'

The so-called pitchstone 'look-alikes' form a small group of materials which are frequently mistaken for volcanic glass, or vice versa. They are:

- Black chert (Plate 4);
- Dark-brown to black homogeneous flint (Plate 5);
- Materials of the jet family (Plate 6);
- Glassy slag (Plate 7);
- Smoky quartz (Plate 8).

Obviously, raw material misidentification is most common in connection with the characterization of smaller artefacts, such as chips. It is equally obvious that certain misidentifications are more likely in certain geographical/ geological areas, such as, 1) where black chert is found naturally and in abundance (eg, southern Scotland, parts of northern England, and Ireland), 2) the core areas of homogenous, highly vitreous flint (eg, parts of eastern and south-eastern England), 3) where jet and jet-related materials are common (eg, around Whitby in Yorkshire), and 4) in areas of prehistoric/historical industrial activities which could cause the formation of glassy slags (eg, ceramic industries); smoky quartz is a relatively uncommon material in British archaeological contexts (but see below).

During the inspection of the lithic assemblages of Scottish museums, several pieces were found which had been characterized as pitchstone, but which were actually homogeneous black chert and, on a number of occasions, pitchstone artefacts were discovered amongst pieces characterized as black chert. The main difference between the two materials is not so much colour as lustre. Although the lustre of both materials can be characterized as 'greasy', that of pitchstone is clearly the more vitreous of the two, as would be expected of a glass. Generally, safe distinction between the two raw materials is a matter of experience and detailed knowledge of the attributes of both: most cherts will include characteristic patterns in the form of little dots, lines or fossil inclusions and, apart from the purest aphyric glasses, most pitchstones include macroscopically visible crystallites and, in some cases, spherulites and phenocrysts (Chapter 3.1).

The highly vitreous dark flints, which in Scotland are most likely to be the purer of the two main forms of imported Yorkshire flint (Durden 1995, 410; Ballin forthcoming i), have a lustre which approaches that of pitchstone. This is particularly noticeable in cases where flint artefacts have developed a secondary sheen. However, in most cases, and with experience, differences can be detected: even if it is not possible to distinguish between the main flint and pitchstone matrices, many flints (including the purest forms) include minuscule chalk spots or fossils, which would not be present in pitchstone.

The problem of mistaking jet, or jet-related materials, for pitchstone is usually only experienced in connection with the characterization of small fragments. In handling larger pieces of jet, it is clearly noticeable that it is much lighter in weight than volcanic glass. When attempting to distinguish between small fragments of jet or pitchstone, lustre is an important attribute. However, as both have a vitreous lustre,

it is again a matter of experience to distinguish between a 'coal-like vitreous lustre' or a 'tar-like vitreous lustre'. Apart from jet proper, which is massive in composition (ie, not grainy or foliated), most other jet-like materials are more or less foliated (Watts & Pollard 1998), which will show in magnification.

Glassy slag may occur in different colours, most commonly green and black. In some cases, the black variety is surprisingly difficult to distinguish from pitchstone, but a small number of slag attributes are not encountered amongst pitchstones. Although, being glassy, this form of slag has a lustre almost identical to that of pitchstone, most pieces are so black that they may appear almost blue, and when light is reflected in their surfaces an iridescent effect may be seen, not dissimilar to that of petrol. If a piece of slag has surviving 'cortex', this will be defined by the material the molten glassy slag had contact with, such as a furnace. And in many instances, pieces of slag may include small surviving air-bubbles or actual impurities.

Smoky quartz is only rarely found on Scottish archaeological sites, but it does occur. In connection with the examination of the Biggar pitchstone collection (Ballin & Ward 2008), a number of dark objects had been identified as pitchstone, and as they were classified as an unfinished microlith, a refitting microburin, and a backed bladelet, it was initially believed that these pieces might represent the first certain Mesolithic pitchstone artefacts encountered outside Arran. However, closer scrutiny revealed that they were all in smoky quartz, and Mesolithic pitchstone artefacts are therefore still absent from the Scottish mainland. The most important point, when attempting to distinguish between pitchstone and smoky quartz, is probably the fact that pitchstone is *translucent*, whereas thin pieces of smoky quartz may be *transparent*.

At present, the greatest problem regarding the potential misidentification of pitchstone is the question of 'English pitchstone': Was pitchstone 'traded' into England or was it not? Figures 24 and 25 clearly show a number of distinct pitchstone concentrations near the Anglo-Scottish border, and a small number of finds south of the border indicate that the exchange network responsible for the dispersal of Arran pitchstone included at least the northern parts of England. If we consider the fact that, in Scotland, archaeological pitchstone has been found 400 km north of the Arran outcrops, it should be possible to recover the occasional piece of pitchstone at least as far south as Manchester. If we also consider the fact that the northwards dispersal of pitchstone probably stops where it does, due to the barrier created by the Atlantic Ocean, worked pitchstone could hypothetically have travelled as far south as the English Channel.

Members of the Implement Petrology Group, and other colleagues south of the Anglo-Scottish border, are assisting with this problem by looking for pitchstone or pitchstone artefacts in their respective regions. Most likely, pitchstone (which is usually black or very dark green) has been misidentified as dark flint (NE England) or black chert (NW England), although smaller fragments may have been characterized as glassy slag or jet.

3.2.2 Altered Pitchstone

The colour of archaeological pitchstone is occasionally puzzling, as it commonly deviates from that of geological pitchstone. Most Scottish collections of archaeological pitchstone include a small number of pieces of light-green or light-brown colours (Plates 9-12). No similar pieces were found during the recent survey of Arran pitchstone outcrops, or during Ballin & Faithfull's (forthcoming) examination of pitchstone samples in the geological stores of the Hunterian Museum and Art Gallery in Glasgow, and it is thought that these pieces may represent pitchstone artefacts which were exposed to fire.

It is well-known amongst lithics specialists that flint reacts to fire in three ways, namely 1) by altering its colour, 2) by developing a degree of crazing, and 3) by loosing weight (Fischer *et al.* 1979, 22). The fact that many light-green and light-brown pitchstone artefacts display crazing, as well as a markedly lower specific gravity than dark-green or black pieces, supports the suggestion that these specimens were exposed to fire. Although some severely burnt pieces, such as those from Auchrannie on Arran (Ballin 2008b) show how hard burning makes pitchstone expand and develop deep fissures (just like heavily burnt flint; Fischer *et al.* 1979, Fig. 10; also, Steffen 2005), lightly burnt pitchstone appears to develop less crazing than flint. Where burnt flint frequently displays marked, macroscopically visible crazing, and takes on an appearance like old porcelain, burnt pitchstone usually displays a form of micro-crazing, which is best seen in magnification. In some instances, this micro-crazing of pitchstone artefacts causes thin edges to crumble (Plate 10).

The most severely burnt pieces of pitchstone tend to display a combination of these different attributes. Like flint, they may have turned completely white (Plate 12); these pieces frequently show the highest degree of weight loss, and many display areas where the surface is turning into fine powder, probably as an extreme result of the disintegrating effect of micro-crazing. The assemblage from Lussa Wood on Jura (Mercer 1980), includes 11 pieces of pitchstone (of a total of 67) which are completely white, exceedingly light in weight, and in the process of turning into powder.

One of the most interesting collections, in connection with the discussion of potentially burnt pitchstone, is the assemblage from Torrs Warren (Cowie 1996), in the Glenluce Sands, Dumfries & Galloway, which includes 179 pieces of pitchstone. Only five of these pieces are black, with the remainder being light greyish-green to light-brown (Plate 13). The vast majority of these generally unworked,

tabular pieces are distinctly lighter in weight and colour than fresh pitchstone, and most are either crazed, or split, with the split usually initiating from a small crystal or spherulite, in the same manner as many thermal flint flakes split due to the presence of a small impurity or fossil (Plate 13). Almost exact formal replicas of the Torrs Warren pieces have been produced in connection with Steffen's (2005) experiments with the burning of obsidian from the Capulin Quarry in New Mexico (Plates 14-15).

In connection with many obsidian assemblages, for example in the United States, burnt obsidian artefacts have been found to display a varying combination of attributes, such as matte finish, surface sheen, fine crazing, deep surface cracking, vesiculation, incipient bubbles, and fire fracture (Steffen 2005, 56). Although some of those have been observed in connection with burnt pitchstone (for example, alteration of finish/sheen and various forms of crazing/ cracking/fracturing, see above), attributes like vesiculation and incipient bubbles have not been recognized yet. The author has been able to replicate vesiculation/incipient bubbles in pitchstone by placing samples in a wood-burning stove in direct contact with fire. This demonstrates that the absence of these attributes in archaeological assemblages is not a matter of differences between the two related raw materials, pitchstone and obsidian – it is more likely a question of differences in pre-deposition activities. It may be that the American samples were exposed to higher temperatures, for example in connection with the deliberate destruction of artefacts, whereas the destruction by fire of pitchstone artefacts was not commonplace in prehistoric Scotland (although the pitchstone from Torrs Warren poses an interesting and somewhat enigmatic case).

In some cases, the exposure of pitchstone/obsidian to fire enhanced faint banding. This phenomenon is covered by Steffen's term 'altered surface sheen' and it is due to variability within the glass (Steffen 2005, 60). This can clearly be seen in Plate 13, where several pieces from Torrs Warren display marked banding. The burnt modified blade from Barnhouse (Plate 16) shows fire-enhanced banding, with the bands running approximately parallel with the long axis of the implement. This attribute was also successfully replicated when the author deposited pitchstone samples in a wood-burning stove.

Several of the Scottish museum collections include light-grey pieces, such as for example the large collection from Biggar in South Lanarkshire (Plate 17; also, Ballin & Ward 2008). This colour is clearly superficial, as indicated by a number of pieces with nicked edges or corners, where the original black or dark green colour is visible, usually immediately beneath the surface. These grey pitchstone artefacts generally have a weathered, slightly abraded appearance, which may represent superficial devitrification and leaching (pers.comm. Dr John Faithfull, Hunterian Museum and Art Gallery, University of Glasgow). These pieces are commonly found in ploughsoil, and the process may correspond to the development of new cortex on flint artefacts, as described by Shepherd (1972).

4. The Evidence – Basic Characterization of Archaeological Pitchstone in Northern Britain

4. 1 Introduction to the Database

As the project's interpretations were to be based predominantly on its pitchstone database, supplemented by information from archaeological literature and from the archaeological community, considerable time was invested in the construction, testing and adjustment of this 'information storage facility'. The most important requirements were:

- The database should be easy to use, that is, it should be simple to store data and to manipulate the information;
- It should be easy to get an overview of the database contents;
- It should be as detailed as possible (that is, within a practical framework), regarding information on typo-technological details and raw material varieties;
- It should be as complete as possible, in terms of known pitchstone-bearing locations;
- It should include references to relevant archaeological literature.

4.1.1 Handling and General Structure

In principle, a database is simply a receptacle for information, and it may be constructed in a multitude of ways. If this receptacle is to include a relatively large amount of information, and if it is thought that relatively sophisticated manipulation of this information may be necessary to allow interpretation of the available data, then the only sensible approach is to construct a 'proper' database. As the most commonly used database programme in the British archaeological community is Microsoft Access, it was chosen to produce the project's database by application of this particular software. An other frequently used programme for data storage is Microsoft Excel, which might have been simpler (or at least equally simple) to deal with during the project's cataloguing phases (Ch. 1.4), but which would not have allowed the data to be manipulated in the same sophisticated manner.

Initially, a detailed database form was designed (the Main Form; Figure 2), but during cataloguing it became clear that it was necessary to store different data in different ways and a second form was designed (the Supplementary Form; Figure 3). All examined assemblages were catalogued by means of the Main Form, and assemblages which it was not possible to gain access to were catalogued using the Supplementary Form. For obvious reasons, the former includes the most detailed characterization of pitchstone assemblages. All entries of pitchstone-bearing locations are referred to by a unique catalogue number, which is either an M-number (Main Form) or an S-number (Supplementary Form).

4.1.2 Overview

It was thought that the best overview of a given assemblage would be achieved, if that assemblage could be presented in a single form, which would only be as large as the computer screen (thus no 'clicking' or 'scrolling' would be needed). It would have been possible to include more data, or use a larger and more user-friendly script, if the information had been stored in a series of linked forms. However, this approach would have complicated the handling of data for each assemblage, and it would have been difficult to gain an overview of the material's attributes. In general terms, the database forms are a compromise, where overview and degree of detail (see below) were weighed against each other.

4.1.3 Degree of Detail

The level of detail in each form was a compromise between the research questions asked, the time available for cataloguing the data (funding), and the need to retain an overview of the stored data. First and foremost, enough data or detail was needed to allow discussion of the defined questions – if this was not the case, the entire project would have been meaningless.

The Main Form was split into three sections, namely site-/assemblage-specific data, characterization of the typo-technological and raw material attributes of the assemblage, and references/comments. The database fields were designed to allow individual collections to be identified, provenanced (where were they found/where are they stored?), and characterized to a degree that would allow discussion of general questions relating to chronology and regionality (including territorial structures and exchange networks).

Due to the size of the form, several groups of non-essential data were excluded, such as radiocarbon dates. Inclusion of

Scottish Archaeological Pitchstone Project, examined artefacts

| Cat no | M001 | Museum | National Museum of Scotland, Edinburgh | | Museum ref | X.BH (see below), X.1997.1013-1015 |

Site information

Provenance	Glenluce Sands		Site type	Stray find(s)
Region	Dumfries and Galloway		NGR	NX 1 5
Date	Early Neolithic		Dating method	Technological profiling

Composition of pitchstone assemblage

| Total lithic assemblage | Thousands | | Total no pitchstone | 190 |

(Main categories)	Unw nodules	3	Debitage	130	Prep flakes	6	Cores	40	Tools	11
(Debitage)	Chips	2	Flakes	28	Blades	66	Microblades	32	Indet pieces	2
(Preparation flakes)	Crested pieces	5	Platf rejuv flakes	1						
(Cores)	Struck nodules	2	Core rough-outs	0	Single-platf	20	Opp-platf	2	2 platf angle	5
	Disc cores	7	Irregular	3	Bipolar cores	0	Other cores	1		
(Tools)	Arrowheads	0	Knives	0	Scrapers	1	Piercers	0	Burins	0
	Truncations	0	Notch/dentic	2	Serrations	0	W inv ret	1	W edge-ret	7
	Other tools	0								
(Pitchstone types)	Aphyric	190	Porphyritic	0						

| Other details | Black, mostly somewhat abraded from deposition in sand. |

| Reference | Thorpe & Thorpe CAT 47-9. |

| Comments | Other core = either irregular core or burin. Scraper = end-scraper on small single-platform core. Museum references: X.BH 8009-8137, 8951, 8960, 9072, 9120, 9151, 9195. |

| Distance to Dun Fionn outcrop (NS 041 339), km | 81 |

Figure 2. Main database form: examined pitchstone artefacts.

Scottish Archaeological Pitchstone Project, un-examined artefacts

This database also includes finds from northern England, Northern Ireland and the Isle of Man

| Cat no | S001 | *Site information* |

| Provenance | East Bennan, Bennan Head | | NGR | NR 9935 2075 |
| Region | Arran | | Assumed date | Early Neolithic |

Assemblage information

| Total no pitchstone | 1 |
| Present location | Unknown. Lost acc. to CANMORE | | Finder/excavator | T.H. Bryce |

| Description of assemblage | |
| 1 'flake of Corriegills pitchstone'. |

| Why not examined | 'Old' assemblage - it has not been possible to locate the finds |

| References | Bryce 1909; Henshall 1972. Thorpe & Thorpe CAT 32. |

| Comments | From burial chamber. Ass. w. Carinated Pottery. |

| Distance to Dun Fionn outcrop (NS 041 339), km | 14 |

Figure 3. Supplementary database form: un-examined pitchstone artefacts.

radiocarbon dates would have been useful, but only if the available space had allowed discussion of the individual dates, their contexts, and the relevance of those dates to the pitchstone artefact(s). However, published radiocarbon dates can be accessed via the assemblage-specific literature listed in the individual entry's 'Reference' field, or via Historic Scotland's Web-site: http://www.historic-scotland. gov.uk/index.htm.

After completion of the cataloguing sections, a final field ('Distance to Dun Fionn outcrop') was defined and inserted. This field was deemed essential to the production of fall-off curves for discussion of exchange and teritoriality. The Dun Fionn source was selected, as it is situated at the centre of the wider Corriegills area (ie, the area in which most aphyric pitchstone is found). Although some porphyritic pitchstone was 'exported' from the Isle of Arran, most exchanged pitchstone was aphyric, and the most homogeneous pitchstone (with the best flaking properties) arguably derives from the Dun Fionn sill, or from smaller outcrops in its immediate vicinity. However, the selection of any site on Arran for this calculation, and subsequently the production of the project's fall-off curves, would have been acceptable, as the distances within Arran are so small compared to the distances over which pitchstone was exchanged (Dun Fionn, Arran – Barnhouse, Orkney = *c.* 400 km).

As the Supplementary Form was defined to store information on un-examined assemblages, it is obviously less detailed than the Main Form, and it provides limited characterization of the finds .

The dimensions of the examined finds (approximately 5,600 individual pieces) were not measured, as this would have made the database substantially more complicated (see above), and it would have required an unacceptable investment of time. However, the assemblage entries do include information relating to artefact size, namely the relative numbers of blades and microblades. The blade:microblade ratio of chronologically unmixed assemblages have been proven to be of value to the relative dating of these collections (eg, Ballin & Ward 2008), and in the present case this information is of relevance to the understanding of technological developments over time, not least in connection with the various pitchstone forms exploited (for more detail on these issues, see Chapters 5 and 6).

4.1.4 Completeness

In connection with a project like the SAPP, it is always difficult to know when to stop: when is the database extensive enough to allow the analyst to answer the questions the database was designed to deal with? The project's focus was on museum collections, and the three cataloguing phases (Chapter 1.4) attempted to deal with the collections of both the region's larger museums and the

smaller ones. However, it became clear that this was not sufficient, as this approach would have skewed the results of the interpretation.

Two major issues were that 1) most museum collections include large amounts of material acquired from field-walkers (ie, biased towards larger and visually more impressive objects, and lacking acceptable information on provenance), and 2) as some parts of northern Britain have only recently experienced more intense development of, for example, infra-structure, several counties had yielded fewer pitchstone finds than one would have expected (based on finds in the surrounding counties), such as the Ayrshire counties.

It was therefore decided to include in the database finds made by amateur archaeologists, as well as finds still held by the excavating units, and a large number of organizations, institutions and individuals were contacted or visited. By combining as yet unregistered finds in the care of National Museums Scotland in Edinburgh with recently excavated finds held by Glasgow University Archaeological Research Division (GUARD), it was possible to add sufficient pitchstone artefacts to the already known artefacts from North, South and East Ayrshire. The distribution of pitchstone on the Scottish mainland east of Arran now appears more representative (see Chapter 7).

The distribution of Arran pitchstone outside present-day Scotland is obviously of importance to the understanding of prehistoric exchange in northern Britain. Consequently, finds from Northern Ireland, the Isle of Man and northern England have also been included in the database (the Supplementary Form).

Readers, who compare the project's database entries with the catalogue entries of Williams Thorpe & Thorpe (1984), will notice a number of omissions (Williams Thorpe & Thorpe CAT 12A, 19, 21, 25, 30, 33, 52, 71, 97, and 100). In the majority of these cases, the existence of these pieces could not be verified, whereas two cases represent double registration (CAT 12, Michael's Grave = CAT 12A, Kilmichael Cairn; CAT 97, Tweed Basin, is simply a general reference to assemblages recovered in the Tweed Valley), and two were in other materials than pitchstone (CAT 33, Monkton, is coal, and CAT 71, Aucrum, is glassy slag).

4.1.5 References

Some of the database entries are not associated with references to the archaeological literature, as those assemblages remain unpublished. Most of these assemblages are relatively old (pre *Discovery and Excavation in Scotland*), small, and unprovenanced/ poorly provenanced (not warranting publication). In most other cases, it has been possible to insert references either to primary excavation reports (mostly in *Proceedings of the Society of Antiquaries of Scotland*), synthetic papers

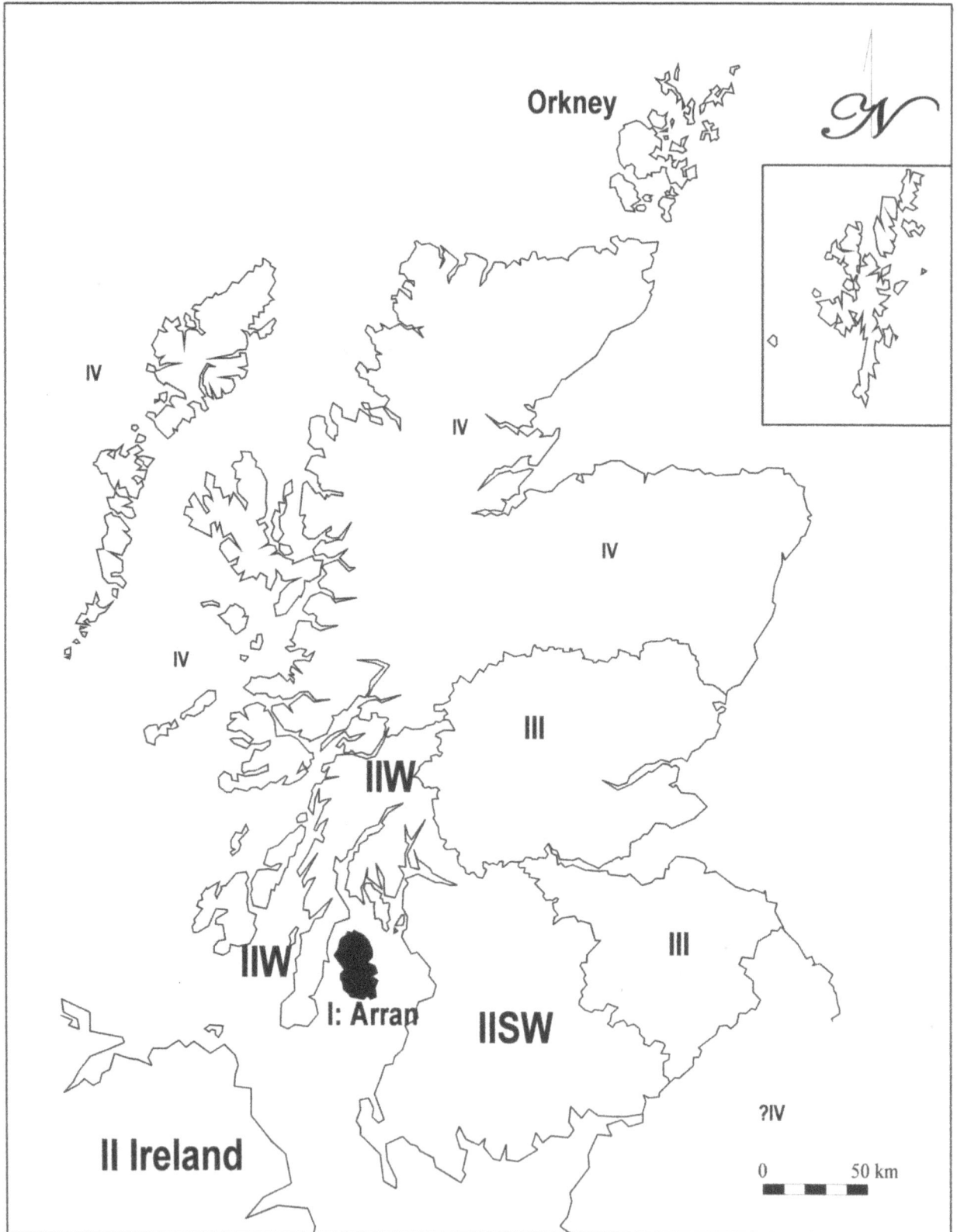

Figure 4. Zone definitions.

Zones	Sub-zones	Counties
I		Arran
II	**W**	Argyll & Bute, incl. Southern Hebrides (from Skye to Islay)
II	**SW**	S/N/E Ayrshire, Inverclyde, Renfrewshire, E Renfrewshire, Glasgow, W/E Dunbartonshire, Dumfries & Galloway, South Lanarkshire, North Lanarkshire
III		Stirling, Clackmannanshire, Perth & Kinross, Fife, Dundee, Angus, E/W/M Lothian, Edinburgh, Falkirk, Scottish Borders
IV		Aberdeenshire, Aberdeen, Moray, Highland, Western Isles
?	**Orkney**	

Table 1. Zone definitions.

(for example, Ritchie 1968; Mulholland 1970; Ness & Ward 2001), or initial brief presentations in *Discovery and Excavation in Scotland*. All entries characterizing assemblages from Williams Thorpe & Thorpe's catalogue have had references to that catalogue included.

Due to the limited space available to each entry, references are given in the shortest possible form. References to actual reports include only author, year, title of periodical and volume (eg, Atkinson 2002 (PSAS 132)). References to entries in *Discovery and Excavation in Scotland* only include title of periodical and volume (eg, DES 1987).

4.2 Scottish Archaeological Pitchstone – General Characterization of the Archaeological Evidence

4.2.1 Introduction

As described in Chapter 4.1, the complete project database includes two sub-databases, one for pitchstone assemblages examined by the analyst, and one for pitchstone artefacts (mostly recently recovered specimens) which were not available for examination. In Chapter 4.2, the finds from the Main Database are presented and characterized in detail. Finds from both databases are used in the discussion of pitchstone distribution (Chapter 7).

This presentation is based on the database as of 31 March 2008. After this cut off point, new pitchstone finds were added to the database, but the new information was not integrated into the report's tables, figures and diagrams. However, these additional finds were added to the Chapter 7 location maps, where they were thought to be of relevance to the understanding of the prehistoric dispersal of pitchstone.

In this and following chapters, northern Britain has been divided into a number of zones (cf. Chapter 1.3). They are defined as in Table 1 and Figure 4.

4.2.2 Present Location of the Finds

All the examined finds were, at the time of writing, in the care of a number of museums, commercial units,

researchers and private collectors. Table 2 indicates where pitchstone artefacts are stored, in terms of numbers of assemblages and individual artefacts. In this context, 'assemblage' is synonymous with an entry in the SAPP database. Finds likely to have been deposited at the same time in prehistory (that is, as a single event) may have been recovered at different times, by different people/organizations, and they may have been recorded individually in, for example, a museum's accession lists. For curatorial reasons, these finds have been entered into the SAPP database in the same manner, that is, to make it easier for researchers to re-trace the artefacts. In Table 2, 'holders' of pitchstone have been sequenced according to whether they belong to one

	Assemblages / database entries	Numbers
National Museums Scotland	96	1,839
Kelvingrove Museum	71	1,208
Biggar Museum	50	692
Dumfries Museum	8	691
Hunterian Museum	60	332
Bute Museum	5	244
Arran Museum	19	118
Stranraer Museum	7	72
McLean Museum	1	35
Orkney Museum	3	26
Marischall Museum	5	21
Historic Scotland	2	12
Paisley Museum	1	5
Perth Museum	2	2
RCAHSM	1	1
Unit: CFA	6	64
Unit: GUARD	10	53
Unit: Headland	3	7
Unit: SUAT	1	1
Others: analysts	2	282
Others: private finders/ collectors	4	32
TOTAL	357	5,737

Table 2. 'Holders' of archaeological pitchstone, assemblages/ entries, and numbers of pitchstone artefacts.

or the other of the categories 'archaeological institutions', 'archaeological units', and 'others', and then by numbers held.

As shown in Table 2, a total of 328 assemblages (5,285 finds) were in the care of Scottish museums, with Historic Scotland holding two assemblages (12 finds) and the Royal Commission one assemblage (one find). Twenty assemblages (125 finds) were with archaeological units, two assemblages (282 finds) were with analysts, whereas private finders and collectors held four assemblages (32 finds). The four largest pitchstone collections, all embracing more than 500 artefacts each, are those of National Museums Scotland, Kelvingrove Art Gallery and Museum, Biggar Museum, and Dumfries Museum. One very large individual assemblage is presently being studied by analyst Diana Coles, Cambridge, namely that of Knocknabb, Torrs Warren, Dumfries & Galloway (280 pieces).

Although some collections are relatively small they may still be important. The finds in the care of the Hunterian Museum and Art Gallery include a large number of assemblages collected by fieldwalking in the Scottish Borders, without which the distribution pattern would have been seriously biased. The assemblages from Bute Museum are unique, as they – in contrast to the finds from most other zones – are dominated by heavily porphyritic material. Although most of the finds from Arran Museum are generally poorly contexted, they include types in pitchstone not found elsewhere, such as oblique arrowheads and barbed-and-tanged arrowheads. The two assemblages from museums west of Glasgow (Moss, near Irvine, North Ayrshire, and Houston South Mound, Renfrewshire) are important, as they represent the prehistory of a geographical area which is generally relatively poorly covered, probably due to a slightly lower than average level of recent development. And Orkney Museum holds the unique Late Neolithic finds from Barnhouse, immediately east of the Stenness/Brodgar ritual complex. The bulk of the finds in the Marischal Museum are thought to be artefacts traded amongst Victorian antiquarians, most of which probably derive from the dune areas of Luce Bay (these finds are heavily abraded from deposition in sand).

Finds still with the archaeological units were considered important to the discussion on pitchstone distribution. The finds recovered recently by CFA Archaeology, for example, supplement the relatively low number of artefacts recovered from the area around the Firth of Forth and, not least, from Edinburgh itself; and the objects recovered by GUARD have added important details to the picture of prehistoric pitchstone use in the Ayrshire counties, which would otherwise have been almost devoid of archaeological pitchstone.

4.2.3 Assemblage Size

The size of the examined assemblages vary between one

and 661 pieces. However, the largest assemblage, which is one of several fieldwalked assemblages from Luce Bay in Dumfries & Galloway, almost certainly represents a mixture of finds from several different settlements in the area. The two second largest assemblages, those from Machrie Moor I and XI (407 and 554 finds, respectively; Haggarty 1991) are probably mixtures of finds deposited during the Mesolithic, Early Neolithic and Late Neolithic periods, although they are clearly heavily dominated by later material. Knocknabb, which, with its 280 pitchstone artefacts, is the fourth largest assemblage in the SAPP database, may represent a single Early Neolithic settlement.

Blackpark Plantation East on Bute (Ballin *et al.* forthcoming) included at the time of examination 240 pitchstone artefacts, and it is thought that probably 80% of those date to the Late Neolithic period, probably representing the largest unmixed non-Arran pitchstone assemblage in Scotland. Since then, further material has been recovered by fieldwalking and small-scale excavation (Sarah Phillips and Anne Speirs pers. comm.), producing a total assemblage of some 400 pieces. If excavated, this assemblage would probably increase to a thousand pieces or more, thus numerically surpassing even Ballygalley in Northern Ireland (Simpson & Meighan 1999).

In connection with the excavation of locations along the Arran Ring Main Water Pipeline, GUARD excavated a number of very large pitchstone-bearing assemblages (Donnelly & Finlay forthcoming). From the largest of these, Site CTSF, a total of 9,363 were recovered, covering the period from the Late Mesolithic to the Late Neolithic. Even if this massive collection is sub-divided into its constituent chronological parts, some of these parts would be considerably larger than any assemblages retrieved outside Arran.

Table 3 and Figure 5 show the average number of pitchstone artefacts per site per zone. In Figure 5, Arran has been excluded, as the average assemblage site is so much larger than in any of the other zones (Table 3). If Arran had been included, all other trends would have been almost imperceptible in the diagram. The figures are not 'pure', in the sense that they reflect prehistoric reality directly, but are influenced by research-historical factors. For example: 1) prior to the computation of the figures from GUARD's Arran Project, and the inclusion of assemblages like the one from Site CTSF, the average size of Arran assemblages was much lower, although still considerably larger than in any of the surrounding zones (when calculated by the analyst in 2006, the average size of Arran assemblages was 63); 2) the size difference in Table 3 between assemblages in Zones IIW and IISW is probably artificial – the relatively low average number from assemblages in Zone IIW will increase as more finds surface at Blackpark Plantation East, and the relatively large figure of assemblages in Zone IISW is influenced by the large fieldwalked collections from the Luce Bay dune areas; and 3) the finds from Orkney are represented entirely by the finds from Barnhouse and the Ness of Brodgar, and the excavation of more domestic Neolithic sites on Orkney, further away from the Stenness/Brodgar ritual complex, would probably cause the island's fairly high average value to decrease somewhat.

In terms of the appearance of Zones III and IV, the information in Table 3 and Figure 5 may be somewhat deceptive. These two zones seem to be almost identical, with assemblages being of roughly the same size. This, however, is an illusion, as demonstrated by the distribution maps in this volume (Figures 24-25). It is correct that the two zones have approximately the same average assemblage size (three and two pieces per pitchstone-bearing site), but the two zones are also characterized by the following important details:

Where Zone IV includes no assemblages larger than four pieces (stray finds from the dunes at Culbin Sands in Moray), Zone III includes several assemblages which number *c*. 20 pieces (Balfarg Riding School, Fife, and Chapelfield, Stirlingshire);

Where pitchstone-yielding assemblages in Zone IV are few and widely spaced, in Zone III they are numerous and, in particular in the Tweed valley, they form dense clusters of sites, indicating that, if excavations were carried out in these areas, the recovery of relatively large assemblages (although probably not quite as large as in neighbouring South Lanarkshire) is almost certain.

In general, the numbers in Table 3 and Figure 5 confirm that the zonation (Figure 4) suggested in previous publications (eg, Ballin 2007a) may actually – with later adjustments and caveats – reflect some form of prehistoric reality (to be discussed further in Chapter 7).

4.2.4 Raw Material Composition

Although up to 100 pitchstone sources are known from Arran (Ballin & Faithfull forthcoming), and although it may be possible via geochemistry, trace element analysis, and crystallite composition to distinguish between many, if not most, of these (eg, Preston *et al.* 1998; 2002), at present it is only safe to distinguish macroscopically between two main pitchstone forms, namely aphyric (or non-porphyritic) and porphyritic pitchstone. In this publication and its associated database, porphyritic pitchstone is defined as

Zones	Numbers
Arran	230
IIW	14
IISW	33
III	3
IV	2
Orkney	14

Table 3. Average number of pitchstone artefacts per site per zone. In this and the following tables, the zones are sequenced according to their distance from Arran (the probable source of all archaeological pitchstone).

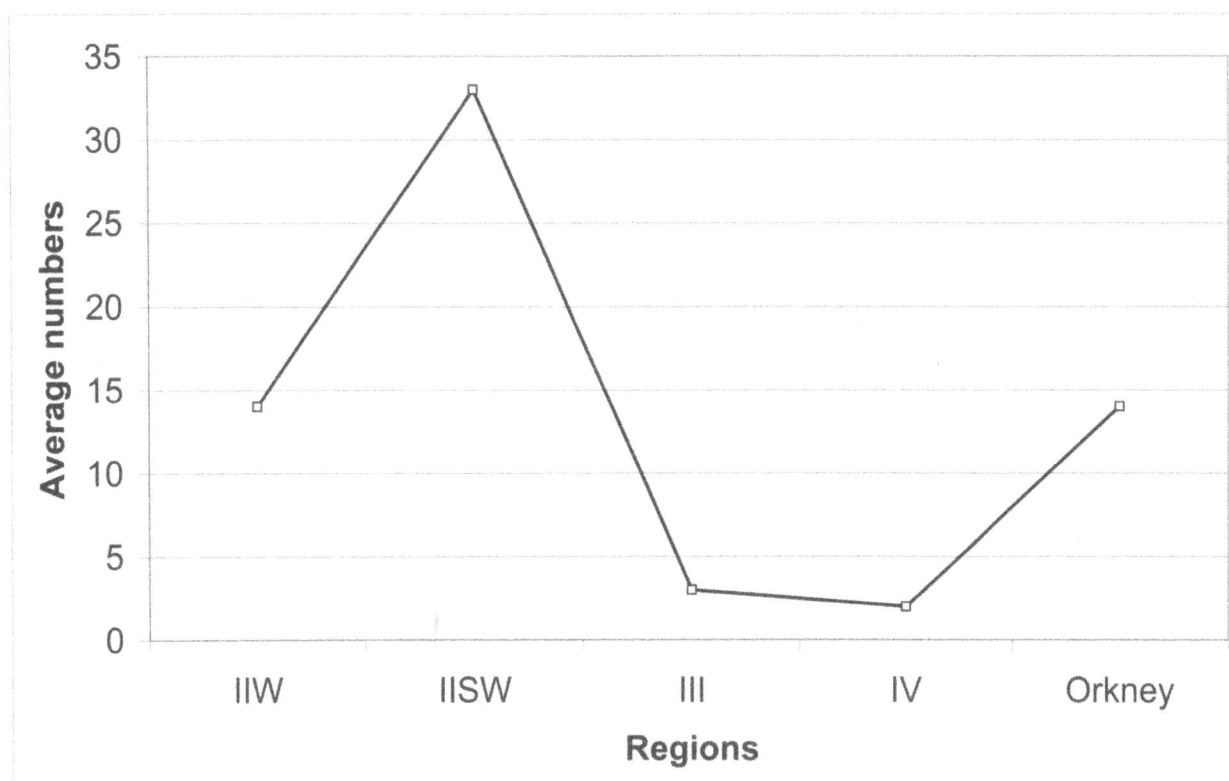

Figure 5. Average number of pitchstone artefacts per site per zone

including *phenocrysts*, that is, smaller or larger crystals, in its glassy matrix, whereas aphyric pitchstone has none. Both pitchstone forms may contain spherulites, most of which are almost microscopic. Spherulites are finely crystalline, usually radiating intergrowths of quartz and feldspar, indicating devitrification of the glass (Chapter 3.1; also, Ballin & Faithfull forthcoming).

As it is not always possible, without the use of microscopic analysis, to distinguish unequivocally between the two forms of pitchstone, these definitions have been qualified in the following way: aphyric, or non-porphyritic, pitchstones may include homogeneous pitchstones with almost microscopic phenocrysts, and porphyritic pitchstones may include aphyric pitchstones with very large spherulites.

Table 4 and Figure 6 show an apparent tendency for sites

to include less and less porphyritic material the farther a location is from Arran, with assemblages from Arran and Zone IIW (Argyll & Bute) including large proportions of porphyritic material (between 40% and 50%), whereas assemblages from Zones IISW, III and IV are almost (but not entirely) devoid of porphyritic material (between 0% and 3%). Orkney (mainly represented by the assemblage from Barnhouse) includes a fairly large proportion (31%) of finely porphyritic material.

It is though that the individual zones' ratios of aphyric:porphyritic pitchstone are the combined effects of chronological, technological, and regional factors (see Chapters 5-7 for detailed discussions). Early Neolithic sites are almost exclusively in aphyric pitchstone (see for example the large assemblages from the Luce Bay area in Dumfries & Galloway, and in the Biggar area in

Zones	Numbers		Total	Per cent		Total
	Aphyric	Porphyritic		Aphyric	Porphyritic	
I	1012	967	*1979*	51	49	*100*
IIW	349	284	*633*	55	45	*100*
IISW	2529	46	*2575*	98	2	*100*
III	248	9	*257*	97	3	*100*
IV	14	0	*14*	100	0	*100*
Orkney	18	8	*26*	69	31	*100*
TOTAL	**4170**	**1314**	*5484*	**76**	**24**	***100***

Table 4. Raw material composition in the various zones.

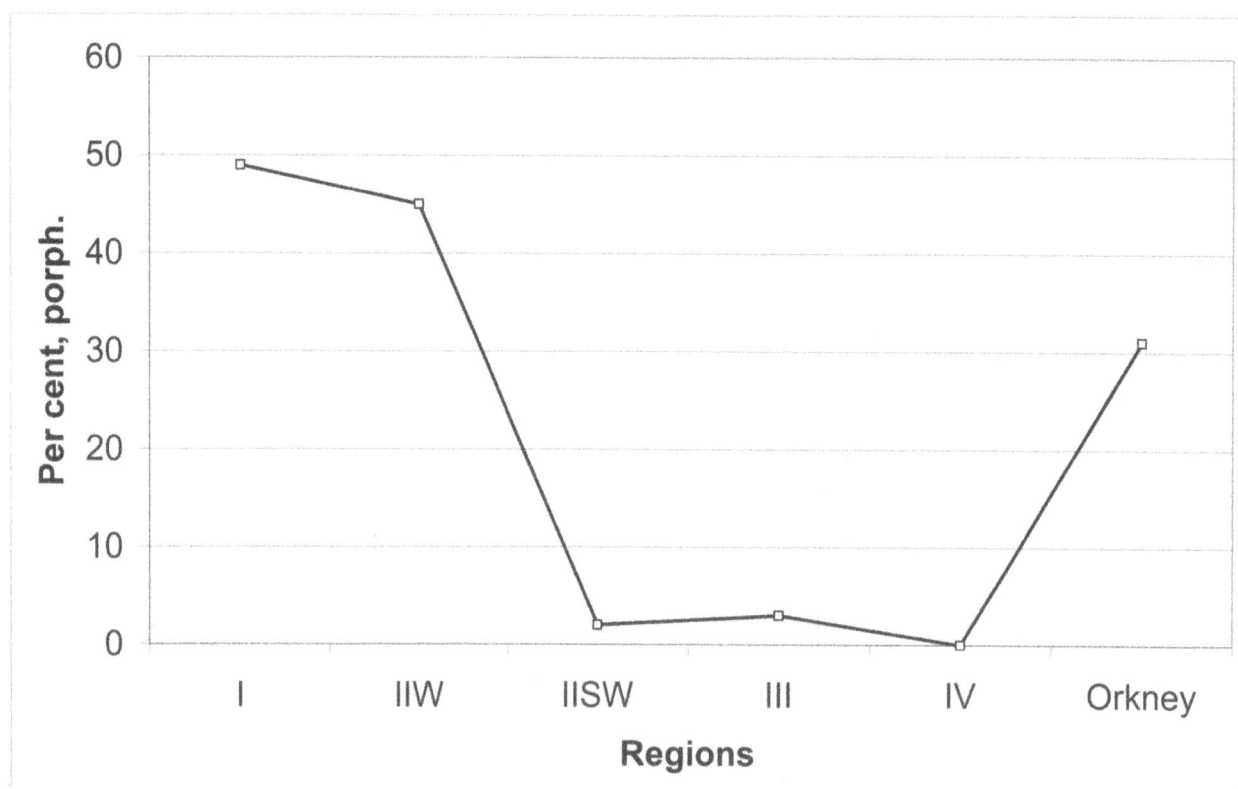

Figure 6. Porphyritic pitchstone as a percentage of all pitchstone, by zone.

South Lanarkshire; SAPP database; Ballin & Ward 2008), with decidedly late assemblages generally including a considerably higher proportion of porphyritic material (eg, Blackpark Plantation East on Bute, and Barnhouse on Orkney; Ballin *et al.* forthcoming; Ballin forthcoming b). However, even Early Neolithic assemblages from Zone IIW include higher ratios of porphyritic material than contemporary assemblages from Zones IISW, III and IV, although generally in forms with very small crystalline inclusions (eg, Auchategan in Argyll and Balloch in Kintyre; Ballin 2006a; Peltenburg 1982; SAPP database).

On occasion, archaeological pitchstone assemblages include light-green/light-brown pieces, as well as grey specimens. It has been suggested that these types of pitchstone may be specific forms of geological pitchstone, but close inspection of pitchstone artefacts in deviating colours suggest that these deviations represent secondary alteration, either in prehistory or after deposition. Most probably, the light-green/light-brown forms are burnt pieces, whereas the grey pieces changed colour either due to contact with soil or air. For a more detail discussion of these phenomena, see Chapter 3.

4.2.5 General Typological Composition (Main Artefact Categories)

Usually, the discussion of the general composition of lithic assemblages includes three categories, namely debitage, cores and tools. In the present case, the assemblages were perceived as including material from five categories, with the additional groups being 'unworked material' (either

	Numbers						Per cent					
Zones	Un-worked	Debi-tage	Prep. flakes	Cores	Tools	Total	Un-worked	Debi-tage	Prep. flakes	Cores	Tools	Total
I	120	1449	11	133	266	1979	6	73	1	7	13	100
IIW	10	527	8	44	44	633	2	83	1	7	7	100
IISW	171	2018	31	217	138	2575	7	79	1	8	5	100
III	1	160	4	21	71	257	1	62	1	8	28	100
IV	1	9	0	1	3	14	7	64	0	7	22	100
Orkney	0	16	1	3	6	26	0	61	4	12	23	100
TOTAL	**303**	**4179**	**55**	**419**	**528**	**5484**	**5**	**76**	**1**	**8**	**10**	**100**

Table 5. General composition of the pitchstone assemblages in the various zones.

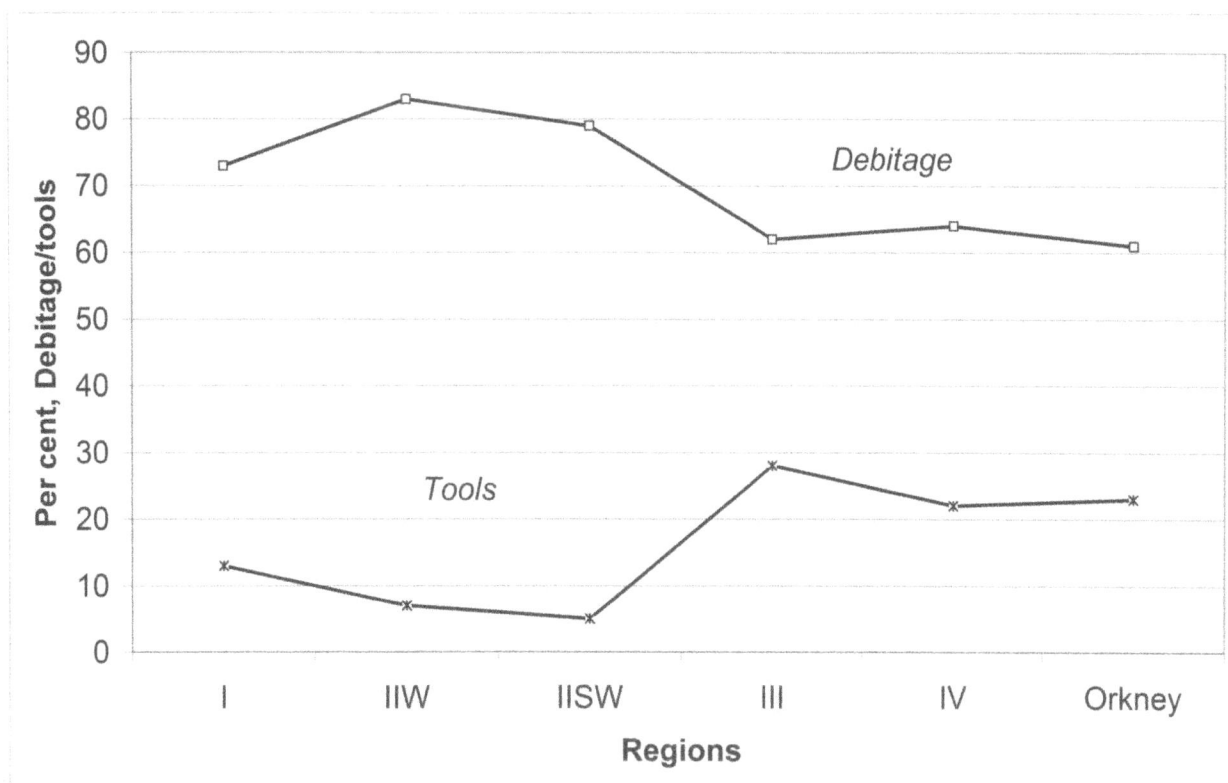

Figure 7. The debitage and tool ratios in the various zones (the core ratios are approximately identical).

21

in the form of tabular pieces or pebbles) and 'preparation flakes' (crested pieces and core rejuvenation flakes). This approach was chosen as, in relation to the non-Arran pitchstone-bearing sites, raw pitchstone would have been as exotic as reduced pitchstone, thereby potentially carrying information of value to the interpretation of the find location, as well as to the discussion of the character of the pitchstone exchange. Preparation flakes were seen as being of potential value to the discussion of whether the recovered pitchstone artefacts had been reduced on the pitchstone-bearing sites or whether they were manufactured on Arran and 'exported' as finished blanks and tools.

Distributed across the various regions, the proportions of the categories 'unworked', 'preparation flakes' and 'cores' do not fluctuate in any systematic way. However, the proportions of the debitage and the tools do. As shown in Table 5 and Figure 7, these two categories co-vary, in the sense that if one category in one region has a high value, the other will automatically have a low value, and vice versa. Basically, tools grow more numerous with growing distance to Arran, whereas unmodified blanks grow less numerous.

Although a myth had developed claiming the opposite ('the further away from Arran an assemblage is found, the fewer tools it includes'), this is not unexpected. A similar phenomenon is known from the Western Isles, where flint is a rare and precious commodity. As a consequence, Western Isles lithics assemblages generally have much higher tool ratios within the flint sub-assemblages than within the sub-assemblages based on the local ubiquitous quartz (at Dalmore on Lewis, for example, flint has a tool ratio of 8% and quartz 1%; at Rosinish on Benbecula, flint has a tool ratio of 61.5% and quartz 0.6%; Ballin 2000; 2002c). If a lithic raw material is precious, either for its rarity or for other inherent values, fewer blanks tend to be left unmodified, abandoned cores may be transformed into tools, and damaged pieces are usually either repaired or recycled.

However, some of the regional tool ratios ought to be considered carefully, as recovery policies are likely to have influenced the ratios somewhat (although not necessarily dramatically so), as for example in the cases of Zone IISW (5%) and Zone III (28%). The finds from Zone IISW include the large assemblages from the Biggar area, much of which has been recovered by excavation, whereas many of the finds from Zone III (practically all finds from the Scottish Borders) have been recovered by fieldwalking. Excavated material generally includes many more chips, which will lower the tool ratio, whereas chips are mostly missed during fieldwalking causing the tool ratio to rise.

4.2.6 Composition of the Individual Main Categories

Unworked material

Although a small number of raw pebbles have been

recovered beyond Arran (such as at Achnahaird Sands in the north-west; Ballin 2002b), unworked pitchstone is usually retrieved in the form of tabular material. The relatively high proportions (6%-7%) of unworked material in some zones off Arran (Table 5) are generally the effects of individual assemblages. In Zone IISW, the high proportion of unworked material is due to the recovery of the Torrs Warren assemblage from Dumfries & Galloway (Cowie 1996), where 83% of an assemblage of 179 pieces of pitchstone is burnt tabular material. In Zone IV, the high proportion is explained by the recovery of one pitchstone pebble at Achnahaird Sands, which in a regional total of 14 pieces of pitchstone counts as 7%. On Arran itself, the high proportion may be more representative of the entire zone, as on the source-island raw pitchstone is readily available.

Debitage

A small number of assemblages are characterized by the presence of relatively large numbers of chips, whereas other assemblages have none. This is most likely a function of differing recovery policies: sieved, or carefully excavated, material may include many chips, where surface collections usually include much smaller numbers, if any.

The differences between flake, blade, and microblade ratios (Table 6 and Figure 8) probably mainly represent different assemblage ages, with Mesolithic and Early Neolithic material including many microblades, later Early and Late Neolithic material some stout blades, and Early Bronze Age material is generally dominated by squat flakes and lacking true blades (Plate 18). At present, assemblages from Arran are influenced by the presence of some fairly large late collections, such as those from the Machrie Moor area (Haggarty 1991; Finlay 1997), but when the huge pitchstone assemblages from recent work on the island (Donnelly & Finlay forthcoming) have been processed and published, the various proportions in Table 6 are likely to change and the proportions of blades and microblades are expected to rise. However, due to the use of pitchstone on Arran after the Early Neolithic period (see Chapters 5 and 6), the island will probably retain a significant proportion of flakes.

The indeterminate pieces of some assemblages may be explained by chronological and raw material differences, with later material containing higher numbers of indeterminate pieces as a result of 1) a less well-controlled reduction strategy (cf. Ballin 2002a), and 2) a larger proportion of porphyritic pitchstone which may cause pieces to disintegrate during reduction (see for example the finds from Blackpark Plantation East on Bute, which is dominated by heavily porphyritic material).

Preparation flakes

Generally, preparation flakes in pitchstone tend to be crested pieces rather than core rejuvenation flakes. In Table 5, of 55 preparation flakes, only eight pieces belong to the latter category. This suggests that initial core preparation took place (cresting), but that core preparation between the individual blank series may have been a less common occurrence (core tablets). The latter may be an effect of the raw material's general attributes, such as the fact that it was provided in the form of relatively *small* tabular pieces. Most likely, the resulting diminutive cores were spent fairly quickly, and discarded after only one or two blank series, thus making platform rejuvenation less relevant. For further details on technological aspects, see Chapter 6.

Cores

The composition of the cores (Table 7) is probably largely a function of site location (that is, distance to Arran) and chronology. Struck nodules and core rough-outs are exclusively found in Zones I-III, and it is thought that this

distribution pattern may be explained in the same way as the regions' tool ratio (see above): the further away from the pitchstone source (Arran), the less likely this treasured resource is to remain unused.

The individual assemblages and zones include different proportions of several characteristic core types, which may largely be due to chronological differences. The most common type is clearly the single-platform core (Plate 19), with most single-platform cores being defined by a more or less flat, untouched 'back-side' (usually the raw surface of a fault line) and a highly regular flaking front, dominated by narrow microblade scars. This fact is probably a consequence of most pitchstone-bearing assemblages off Arran dating to the Early Neolithic period (see Chapter 5). The opposed-platform cores, cores with two platforms at an angle, and irregular cores may mainly be later stages of the

	Numbers						Per cent					
Zones	Chips	Flakes	Blades	Micro-blades	Indet. pieces	Total	Zones	Chips	Flakes	Blades	Micro-blades	Indet. pieces	Total
I	155	996	95	51	152	1449	I	11	69	7	3	10	100
IIW	65	294	72	62	34	527	IIW	12	56	14	12	6	100
IISW	205	703	470	592	48	2018	IISW	10	35	23	29	3	100
III	3	64	61	31	1	160	III	2	40	38	19	1	100
IV	0	5	1	2	1	9	IV	0	56	11	22	11	100
Orkney	2	13	1	0	0	16	Orkney	13	81	6	0	0	100
TOTAL	430	2075	700	738	236	4179	TOTAL	10	50	17	18	5	100

Table 6. Composition of the debitage in the various zones.

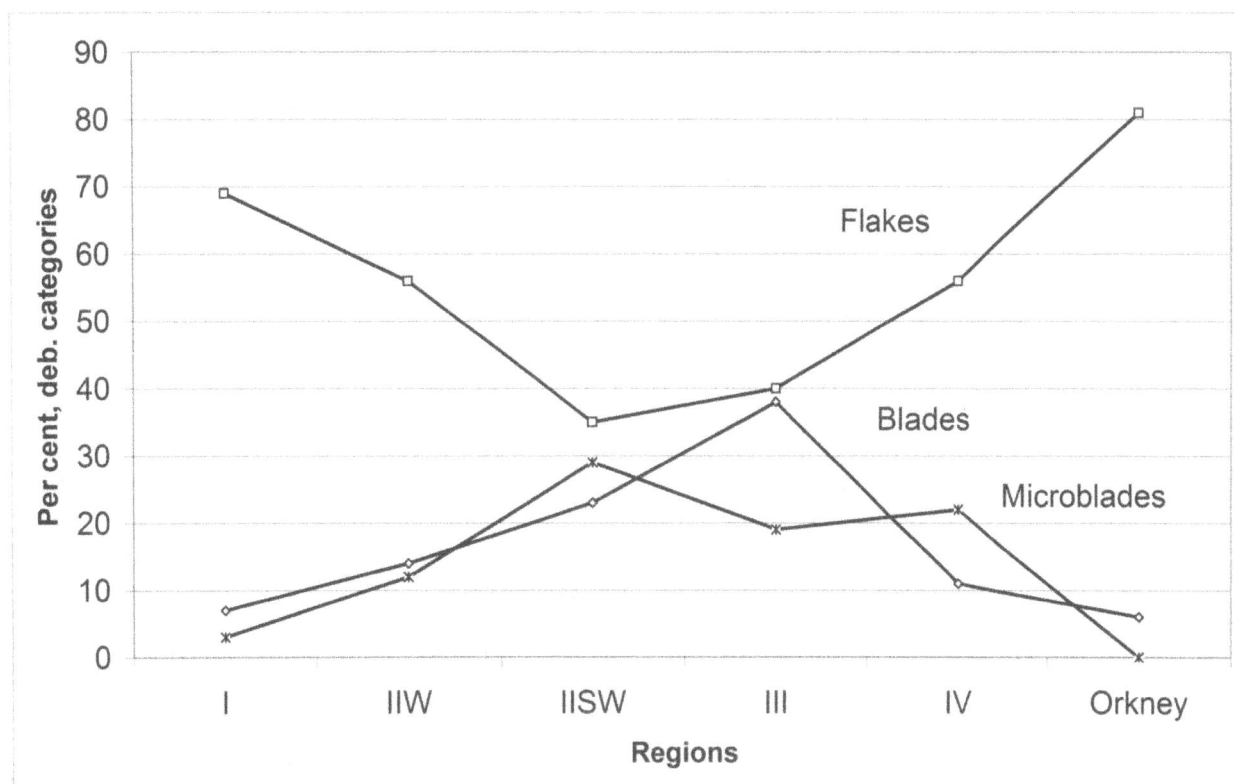

Figure 8. Main debitage categories as a percentage of all debitage, by zone.

Numbers

Zones	Struck nodules	Core rough-outs	Single-platf cores	Opp-platf cores	2 platf at angle	Disc cores	Irregu-lar cores	Bipolar cores	Other cores	Total
I	3	5	54	11	5	6	16	27	6	133
IIW	4	0	18	4	3	2	8	2	3	44
IISW	14	4	90	34	18	20	19	7	11	217
III	0	1	7	1	3	5	2	2	0	21
IV	0	0	1	0	0	0	0	0	0	1
Orkney	0	0	0	0	0	2	0	1	0	3
TOTAL	**21**	**10**	**170**	**50**	**29**	**35**	**45**	**39**	**20**	**419**

Per cent

Zones	Struck nodules	Core rough-outs	Single-platf cores	Opp-platf cores	2 platf at angle	Disc cores	Irregu-lar cores	Bipolar cores	Other cores	Total
I	2	4	40	8	4	5	12	20	5	100
IIW	9	0	40	9	7	5	18	5	7	100
IISW	7	2	41	16	8	9	9	3	5	100
III	0	5	33	5	14	23	10	10	0	100
IV	0	0	100	0	0	0	0	0	0	100
Orkney	0	0	0	0	0	67	0	33	0	100
TOTAL	5	2	41	12	7	8	11	9	5	100

Table 7. Composition of the cores in the various zones.

single-platform cores, representing attempts at exhausting usable raw material. The Machrie Moor material includes some large discoidal cores, several of which belong to the Levallois-like family of cores (Plate 20-21), thus suggesting a Late Neolithic date for parts of these collections (Ballin forthcoming a).

The examination of the rich pitchstone finds from the Glen Luce area in Dumfries & Galloway allowed the definition of a specific form of discoidal core (discoidal cores of 'Glen Luce Type'; Plates 22-23), which is rarely (if ever?) seen in other raw materials. In a sense, this type is a hybrid core form, with elements from discoidal cores and cores with two platforms at an angle. It is most certainly discoidal, in terms of its general shape, but the microblades detached from the two faces are orientated at perpendicular angles to

each other (Figure 9). In contrast to this, most typical cores with two platforms at an angle are rather cubic specimens. It is possible that the creation of this core type is a result of the exaggerated tendency of pitchstone blades to curve along the long axis. This core type is particularly common in Zones IISW and III (Ballin & Ward 2008), and it probably dates to the Early Neolithic period.

Bipolar cores are rarely recovered from pitchstone-bearing assemblages, and it is thought that the use of bipolar technology in connection with pitchstone reduction could be a later occurrence. The assemblage CAT 03, probably recovered from Glenluce Sands, includes 16 pieces of pitchstone, mostly in a deviating variety of pitchstone (slightly coarser than other forms of volcanic glass from the area), and four of the 16 pieces are bipolar cores. No other

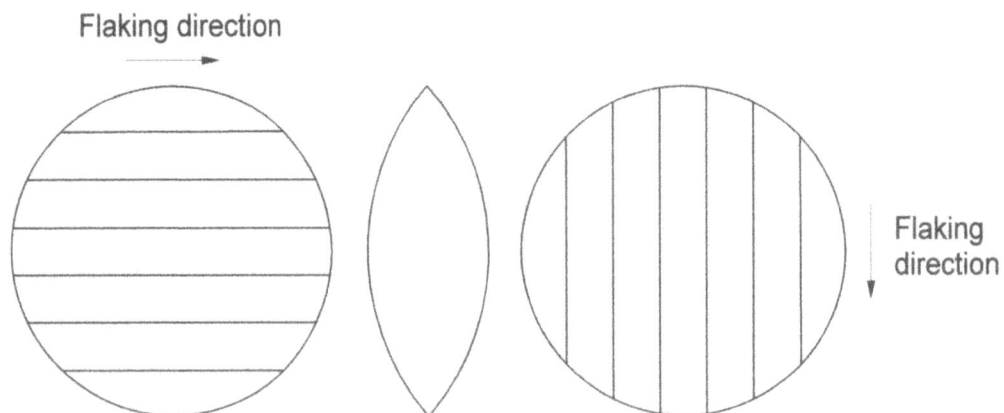

Figure 9. A typical small discoidal core in pitchstone ('Glen Luce Type').

Numbers

Zones	Arrow-heads	Knives	Scra-pers	Pier-cers	Burins	Trunca-tions	Notch/dentic	Serra-tions	W inv ret	W edge-ret	Other tools	Total
I	21	3	61	6	2	7	19	0	8	139	0	266
IIW	1	1	11	0	1	1	5	0	0	24	0	44
IISW	3	2	20	6	1	18	5	0	3	79	1	138
III	1	5	10	1	0	13	5	0	2	34	0	71
IV	0	0	0	2	0	0	0	0	0	1	0	3
Orkney	0	0	2	0	0	0	1	0	0	3	0	6
TOTAL	**26**	**11**	**104**	**15**	**4**	**39**	**35**	**0**	**13**	**280**	**1**	**528**

Per cent

Zones	Arrow-heads	Knives	Scra-pers	Pier-cers	Burins	Trunca-tions	Notch/dentic	Serra-tions	W inv ret	W edge-ret	Other tools	Total
I	8	1	23	2	1	3	7	0	3	52	0	100
IIW	2	2	25	0	2	2	12	0	0	55	0	100
IISW	2	1	15	4	1	13	4	0	2	57	1	100
III	1	7	14	1	0	19	7	0	3	48	0	100
IV	0	0	0	67	0	0	0	0	0	33	0	100
Orkney	0	0	33	0	0	0	17	0	0	50	0	100
TOTAL	**5**	**2**	**20**	**3**	**1**	**7**	**7**	**0**	**2**	**53**	**0**	**100**

Table 8. Composition of the tools in the various zones.

cores were recovered from this site. Throughout Scotland, the average bipolar core ratio is 9%, but this figure varies from region to region. The high ratio encountered on Arran (20%) is probably largely a function of the relatively late date of many Arran assemblages (flake industries), and the low ratio (3%) encountered in Zone IISW may largely be due to the relatively early (Early Neolithic) date of most assemblages. In all other cases, the numbers of bipolar cores and the zone totals are so small, that any deviation from the average may be explained as a statistical, rather than an archaeological, phenomenon.

Tools

The composition of the tool category is probably mainly a result of the economy of the individual assemblages, and assemblage date. No regional trends were identified. As mentioned above, in connection with the discussion of the general composition of the various pitchstone assemblages (Table 5), the relatively large number of tools found on the mainland effectively disposes of the popular myth that pitchstone assemblages from this part of Scotland include almost no tools. If a tool is defined as a secondarily modified blank, off-Arran assemblages include from 5% to 28% tools, and the individual assemblages include specimens of all known tool types, apart from serrated pieces (Table 8).

The most distinctive difference between pitchstone assemblages and, for example, flint assemblages is the fact that the former include substantially higher numbers of simple edge-retouched forms, supplemented by relatively small numbers of formal tools. Edge-retouched flakes and blades generally make up between 33% and 57% of the

tools, where in contemporary flint collections this category usually accounts for much less.

Arrowheads (Plates 24-25) of all types are more common on Arran than outside the island, but arrowheads *are* known from mainland sites: from mainland Scotland, arrowheads have been recovered in connection with the investigation of the Luce Sands area in Dumfries & Galloway (one rough-out for a leaf-shaped arrowhead and one chisel-shaped arrowhead), Scottish Woodlands Area South near Biggar (one chisel-shaped arrowhead), Lealt Bay on Jura (one leaf-shaped arrowhead), and Ardownie Farm in Angus (one leaf-shaped arrowhead). Recently, a weathered leaf-shaped point was recovered during fieldwalking on Kintyre, Argyll & Bute (Vicki Cummings, pers. comm.; not included in Table 8). Presently, oblique arrowheads and barbed-and-tanged arrowheads have only been retrieved from sites on Arran, and microliths *sensu stricto* are also absent outwith this island[1].

All 11 knives are expedient pieces, one of which is on a blade and the remainder on small flakes. Four pieces are plain backed pieces, whereas most are knives with simple invasive or semi-invasive retouch ('scale-flaking') along the cutting-edge. Pieces with truncations (mostly straight and oblique truncations) are generally based on microblades

---

[1] It is proposed to follow Clark's definition of a microlith as *a bladelet which has had its proximal end removed* (Clark 1934a, 55). This definition allows the microlith to maintain its diagnostic value as a pre-Neolithic artefact (cf. Ballin 1996b); the definition also separates microliths 'proper' from the less diagnostic backed bladelets with intact bulbar ends. In Scandinavian typology a microlith is usually defined as a microblade which has had its proximal end removed by the application of microburin technique (Brinch Petersen 1966, 93).

or very narrow blades. Due to their small sizes, they are thought to have served as edges in composite tools. Scrapers in pitchstone (Plate 26) are relatively common throughout Scotland, and traditional piercers (Plate 27) are also occasionally found. In addition, four possible burins have been retrieved, two from Arran and one from Argyll & Bute, which may have formed part of the same social territory as Arran, and where pitchstone may have been used slightly earlier and later than in other parts of off-Arran Scotland. The fourth possible burin, which is held by National Museums Scotland, has been provenanced as 'possibly from Glenluce Sands'. Thirteen fragments with invasive retouch may be parts of post-Mesolithic implements or rough-outs.

Accepting that probably most of the mainland pitchstone finds date to the Early Neolithic period (Chapter 5), it is interesting that no serrated pieces have been recovered, on or outwith Arran. This fact may reflect the relatively brittle, as well as soft, character of the raw material, causing it to splinter relatively easily. Over time, pitchstone cutting-edges (serrated or not) were probably dulled relatively quickly, as demonstrated by the frequently abraded working-edges of pitchstone artefacts, as well as by their scratched surfaces.

5. The Date of Pitchstone Use in Northern Britain

5.1 Introduction

The dating of prehistoric pitchstone use in northern Britain is arguably the most outdated aspect of pitchstone research, in the sense that there is now a vast chasm between general consensus, and the views expressed in the 'classic' pitchstone literature (eg, Mann 1918; Ritchie 1968; Mulholland 1970; Williams Thorpe & Thorpe 1984). As the most exciting questions regarding the prehistoric use of pitchstone (eg, procurement, exchange networks and territories) can only be dealt with on the basis of a sound chronology of pitchstone use, this question is given special attention here. This chapter is divided into two main parts: the first discusses the dating of pitchstone use on the Isle of Arran, and the second, the dating of pitchstone use beyond Arran. As the main focus of the present project is to analyse the distribution of pitchstone outwith Arran, the chronology of the Arran material is presented in summary form, whereas the chronology of the non-Arran material is discussed in greater detail.

In the older archaeological literature, it is generally suggested that pitchstone was used – on and off Arran – through most of prehistory, from the Mesolithic to the Early Bronze Age. This view was mainly based on the erroneous understanding that microblades and narrow macroblades were exclusively Late Mesolithic, as well as the relatively uncritical use of poorly contexted, frequently chronologically mixed, assemblages. The more recent recovery of worked pitchstone from secure and datable contexts, as well as the sheer volume of this material, provides a greatly improved basis for chronological assessment.

In general, archaeological pitchstone can be dated using the following key elements:

- Diagnostic types;
- Diagnostic technological approaches (operational schemas) and attributes;
- Raw material preferences (aphyric/porphyritic pitchstone);
- Find contexts; and
- The association with other datable find groups.

5.2 Arran

Examination of diagnostic tool types found on Arran indicates that here pitchstone was procured for the manufacture of implements throughout most of the stone-tool using part of prehistory. A number of pitchstone microliths have been retrieved (Affleck *et al.* 1988, 47; Donnelly & Finlay forthcoming), as well as leaf-shaped, chisel-shaped, oblique, and barbed-and-tanged arrowheads (Haggarty 1991; Finlay 1997; SAPP database). However, pitchstone microliths are relatively rare, and this fact, in conjunction with a marked dominance of flint on some Mesolithic sites (eg, Auchareoch, where flint dominates pitchstone at a ratio of 9:1; Affleck *et al.* 1988), suggests that pitchstone may have remained a supplementary lithic raw material until the onset of the Neolithic period.

As little Early Mesolithic material is known from Arran, it is not possible to say exactly when the post-glacial settlers started exploiting this resource. However, comparison with similar post-glacial scenarios in other parts of North-West Europe suggests that this may have been a gradual process. In Norway, for example, the first settlers used flint, as this was the material they knew from their origins in southern Scandinavia and on the North-West European plain. As they gradually became acquainted with their new territories, they learnt where alternative resources were to be found, and they discovered how to work these new raw materials, many of which were considerably more abundant than flint (Bruen Olsen 1992, 84; Ballin 2004a; 2007b).

From the beginning of the Neolithic, pitchstone gained in importance, and it soon became the dominant lithic raw material. Although the sites included in Table 9 may be mixed to varying degrees, the assemblages are thought to date largely to the Early Neolithic period, although they also include elements from later parts of the Neolithic (Finlay 1997). The impression of the island's lithic assemblages as being dominated by pitchstone throughout the Early and Late Neolithic periods is supported by the typological and technological characterization of the finds from Machrie Moor I and XI (Haggarty 1991; SAPP database). These two collections contain some Mesolithic and Early Neolithic material, but the relatively large numbers of plain chisel-shaped arrowheads imply dominance of Late Neolithic artefacts. On these sites, chisel-shaped arrowheads in

	Quantity				Per cent			
	Flint	Aphyric pitchstone	Porphyritic pitchstone	Total	Flint	Aphyric pitchstone	Porphyritic pitchstone	Total
Kilpatrick Settlement 16/1	82	159	28	269	30	59	11	100
Kilpatrick Cairn 16/2	38	59	30	127	30	46	24	100
Kilpatrick Cairn 16/3	24	25	5	54	44	46	9	100
Kilpatrick 'Fernie Bank' field boundary	24	67	23	114	21	59	20	100
Machrie Moor Cairn 24/1	33	111	4	148	22	75	3	100
TOTAL / AVERAGE	201	421	90	712	28	59	13	100

Table 9. The raw material distribution of the more substantial assemblages from The Arran Prehistoric Landscape Project (based on information in Finlay 1997).

pitchstone outnumber leaf-shaped arrowheads in pitchstone by a factor of eight to one.

The fact that, at Machrie Moor, the pitchstone microblades and their associated cores are exclusively in aphyric material, whereas the broader blades and the Levallois-like cores are largely in porphyritic material, indicates a shift – at least in the western parts of Arran – in terms of pitchstone preferences. Obviously, aphyric material is best for the production of small, thin bladelets, and it appears that, when those were no longer being produced, local porphyritic pitchstone was being preferred to the aphyric varieties obtained from the island's east-coast. If this impression is correct – on Arran, Early Neolithic assemblages are dominated by aphyric material, whereas Late Neolithic assemblages include much more porphyritic material – this supports the dating of the assemblages in Table 9 as largely Early Neolithic. The Machrie Moor assemblages have porphyritic ratios of approximately 50-60%, whereas the assemblages in Table 9 have ratios of 5-35% (when the flint component is disregarded).

Although oblique and barbed-and-tanged arrowheads in pitchstone are known on Arran, these are slightly less common than earlier arrowhead forms, such as leaf-shaped and chisel-shaped pieces. This suggests that pitchstone may have been in use throughout the Late Neolithic/Early Bronze Age period, but that it gradually lost its importance. At present, few oblique or barbed-and-tanged points are known from excavated settlement sites.

Technologically, the pitchstone finds of Arran can be divided into two main groups: those assemblages based on the production of microblades and narrow macroblades on mainly single-platform cores, and those involving the production of broad blades and elongated flakes on mainly single-platform and discoidal core types. The Machrie Moor assemblages include a number of decidedly Levallois-like discoidal cores (cf. Ballin forthcoming a). The narrow and broad blades are characterized by different sets of technological attributes, with the former largely having plain abraded platform remnants, and the latter finely faceted platform remnants. The beginning and the end

of the Neolithic period are clearly associated with very different operational schemas.

Although the vast majority of Arran pitchstone artefacts have been recovered from settlement sites, not least as part of several large-scale area investigations (eg, Barber 1997; Donnelly & Finlay forthcoming), as well as many smaller development projects, a large number of pitchstone objects have been retrieved from Early Neolithic burial monuments. Most of these cairns were investigated by Bryce at the beginning of the twentieth century (Bryce 1903; 1909), but some have been excavated more recently (eg, MacKie 1964). Although finds from chambered cairns may be difficult to date, due to the general re-use of these communal monuments, the pitchstone artefacts from Arran's chambered cairns clearly post-date the Mesolithic period, and most of them are thought to be Early Neolithic. A total of 258 artefacts from the SAPP's pitchstone database derive from Neolithic burial monuments, one of which is a leaf-shaped point (Blairmore).

5.3 The Situation beyond Arran

As mentioned above, older archaeological literature suggested that beyond Arran pitchstone was used throughout most of the stone-using part of prehistory, from the Mesolithic to the Early Bronze Age. This supposition is discussed below with reference to typological, technological, raw material and contextual evidence, as well as the associations of worked pitchstone with other diagnostic find groups.

5.3.1 Diagnostic Types

In essence, the discussion of the typological dating evidence focuses on the presence or absence of known diagnostic core and tool types. Off Arran, typical Mesolithic types (such as burins, microliths[2], and microburins) are absent in pitchstone collections. However, where some tool types may be reliable chronological indicators in connection with the assessment of, for example, the dating of flint

---

[2] See definition of microliths in Chapter 4.2.6.

or chert assemblages, they are not necessarily reliable diagnostic types in connection with the assessment of pitchstone assemblages. Although burins form an integral part of Scottish Early and Late Mesolithic flint and chert assemblages (eg, Howburn and Glentaggart in South Lanarkshire; Ballin & Johnson 2005; Saville et al. 2008), they are absent in pitchstone collections on the Scottish mainland, and rare in pitchstone assemblages on Arran itself (eg, Donnelly & Finlay forthcoming). This may be due to pitchstone being a relatively brittle material, which is poorly suited for the heavy-duty work usually associated with burins (for the shaping of artefacts in wood, antler, and bone). The brittleness of pitchstone is also indicated by the fact that Neolithic serrated pieces in this raw material are completely absent both on and off Arran, and by the fact that the robust bipolar technique is applied relatively infrequently in connection with the reduction of pitchstone.

Microliths and microburins, on the other hand, are relatively common in Late Mesolithic pitchstone assemblage on Arran (Donnelly & Finlay forthcoming). If the exchange of pitchstone was initiated from the moment the reduction of this raw material first took place on Arran (probably in the Late Mesolithic period), pitchstone microliths and microburins should have been expected to occur on the Scottish mainland. Due to the presence of numerous microblades, it has been assumed that the bulk of the pitchstone artefacts from Luce Bay and Biggar are Mesolithic. However, neither of these two large assemblages includes any microliths or microburins. During the examination of the pitchstone collection from Biggar, several very dark microliths were noted, and in one case it was possible to unite a small dark microlith with its microburin. However, closer scrutiny revealed that none of these pieces was in pitchstone, most being in black chert, and the refitted microlith and microburin were identified as artefacts made of dark smoky quartz.

The known off-Arran pitchstone-bearing sites which are usually considered to be Mesolithic in date (due to the presence of Mesolithic diagnostic types in flint, chert or bloodstone) are not very helpful in this context, as 1) they are either multi-occupation sites with a degree of Neolithic or Bronze Age intrusion (eg, Kinloch on Rhum, several of Mercer's sites on Jura, Ballantrae in South Ayrshire; Wickham-Jones 1990; Mercer 1968; 1971; 1972; 1980; Lacaille 1945), 2) they were recovered by fieldwalking, leaving no possibility of chronological control (eg, Shewalton Moor; Lacaille 1930; 1937), or 3) they were excavated so long ago that their excavation approach is today considered unsatisfactory (eg, Barsalloch in Dumfries & Galloway; Cormack 1970).

A number of post-Mesolithic arrowheads in pitchstone have been found on sites outwith Arran, namely Neolithic leaf-shaped and chisel-shaped specimens. Late Neolithic oblique points and Early Bronze Age barbed-and-tanged arrowheads, which are known from Arran, have not been found off this island. Three leaf-shaped arrowheads are known off Arran: one from Lealt Bay on Jura; one from Luce Bay in Dumfries & Galloway, and one from Ardownie in Angus (SAPP database). Two chisel-shaped arrowheads have been identified, with one each from Luce Bay in Dumfries & Galloway, and from the Scottish Woodlands Area South near Biggar in South Lanarkshire (Williams Thorpe & Thorpe 1984; Ballin & Ward 2008). It seems that, on the Scottish mainland, Late Neolithic chisel-shaped and oblique arrowheads are largely in imported grey Yorkshire flint (Stevenson 1948), whereas pitchstone versions of these points are relatively common on Arran itself (Machrie Moor and stray finds; SAPP database).

In the report (Ballin forthcoming c) on the extensive lithic assemblage from the Raunds Area Project, Northamptonshire, the author discussed the relative dating of later Neolithic chisel-shaped and oblique arrowheads:

'At Hunstanton, a number of chisel-shaped arrowheads were associated with Grooved Ware (Healy 1993, 34), suggesting that the general perception of chisels being associated with Peterborough Ware and oblique arrowheads with Grooved Ware may be an over-simplification (cf. discussion in Saville 1981, 49-50). This notion is further supported by the material from pits at Fengate, where chisel-shaped and oblique arrowheads were found together (eg, Pit W17; Pryor 1978, 21), and Green (1980, 235-6) has documented the association of chisel arrowheads with the Clacton and Woodlands sub-styles'.

In summary: oblique arrowheads are probably associated exclusively with Late Neolithic Grooved Ware contexts, whereas the simpler forms of the transverse arrowhead are associated with slightly earlier Late Neolithic Peterborough Ware / Impressed Ware contexts, as well as with Grooved Ware contexts.

As mentioned above, serrated pieces, which are generally considered diagnostic of Early as well as Late Neolithic assemblages (Saville 2006; Ballin forthcoming g), have not been encountered in pitchstone at all. It is quite symptomatic that, in the collection from Late Neolithic Blackpark Plantation East on Bute (Ballin *et al.* forthcoming), there was a well-executed finely serrated blade in flint but none in pitchstone, which was the collection's dominant raw material. Highly diagnostic Levallois-like cores have only been recovered from sites on Arran (Machrie Moor; SAPP database), and one was found at Barnhouse on Orkney. This type is an unequivocal Late Neolithic indicator (Ballin forthcoming a; forthcoming g).

5.3.2 Diagnostic Technological Attributes

In connection with the dating of pitchstone artefacts, a

number of technological attributes and combinations of attributes are diagnostic to varying degrees, such as:

- Dominant blank type;
- The preparation of the blanks' platform-edge;
- The preparation of the blanks' platform surface; and
- The operational schema as a whole.

As mentioned above, it has been common practice to assign pitchstone assemblages dominated by microblades to the Late Mesolithic period. However, recent research (see for example the radiocarbon dating of pitchstone microblades in pits, below, and the flint assemblage from Garthdee Road in Aberdeen; Ballin forthcoming e) has shown that narrow blades are common in the Early Neolithic period as well. Figure 10 (for details, see Chapter 4.2.6) divides the regions into two groups. The first group includes the south-west, the south-east/east and the Highland region (Zones IISW, III and IV) and the second includes Arran, the west of Scotland and Orkney (Zones I, IIW and Orkney). The former group has a high percentage of microblades (20-33%), whereas the latter has a much lower percentage (0-14%). This probably implies that the former group is dominated by earlier material, whereas the latter includes some later material as well.

However, as also shown in this diagram, the most interesting co-variance is not so much between microblades and blades as between blades/microblades and flakes. The regions with low microblade ratios are also regions with low blade ratios, but high flake ratios. This indicates that the collections from Arran, the west of Scotland and Orkney may include assemblages from a period where blades, as such, were in the process of being phased out, such as the Late Neolithic period and the Early Bronze Age. With microblade ratios of between 20-30%, supplemented by approximately equal numbers of blades (which would usually only be millimetres broader than the microblades), the assemblages from the south-west, the south-east/east and the Highland region are most likely to be dominated by Early Neolithic material.

In connection with the discussion of the finds from Arran, it was mentioned that on this island it was possible to define two distinct operational schemas: the Early Neolithic schema was aimed at the production of microblades and narrow macroblades on mainly conical single-platform cores, whereas the Late Neolithic schema aimed at the production of broad blades and elongated flakes on mainly broad single-platform and Levallois-like core types. The former schema is primarily characterized by blades having plain abraded platform remnants, and the latter by blades having finely faceted platform remnants.

Beyond Arran, practically all pitchstone assemblages are defined by the former operational schema, with blades generally having plain platforms and abraded platform-edges. The only assemblage where the latter operational schema was followed is Barnhouse on Orkney (Ballin forthcoming b), where one Levallois-like core was recovered, and where one flake has the typical finely faceted platform remnant associated with Levallois-like production.

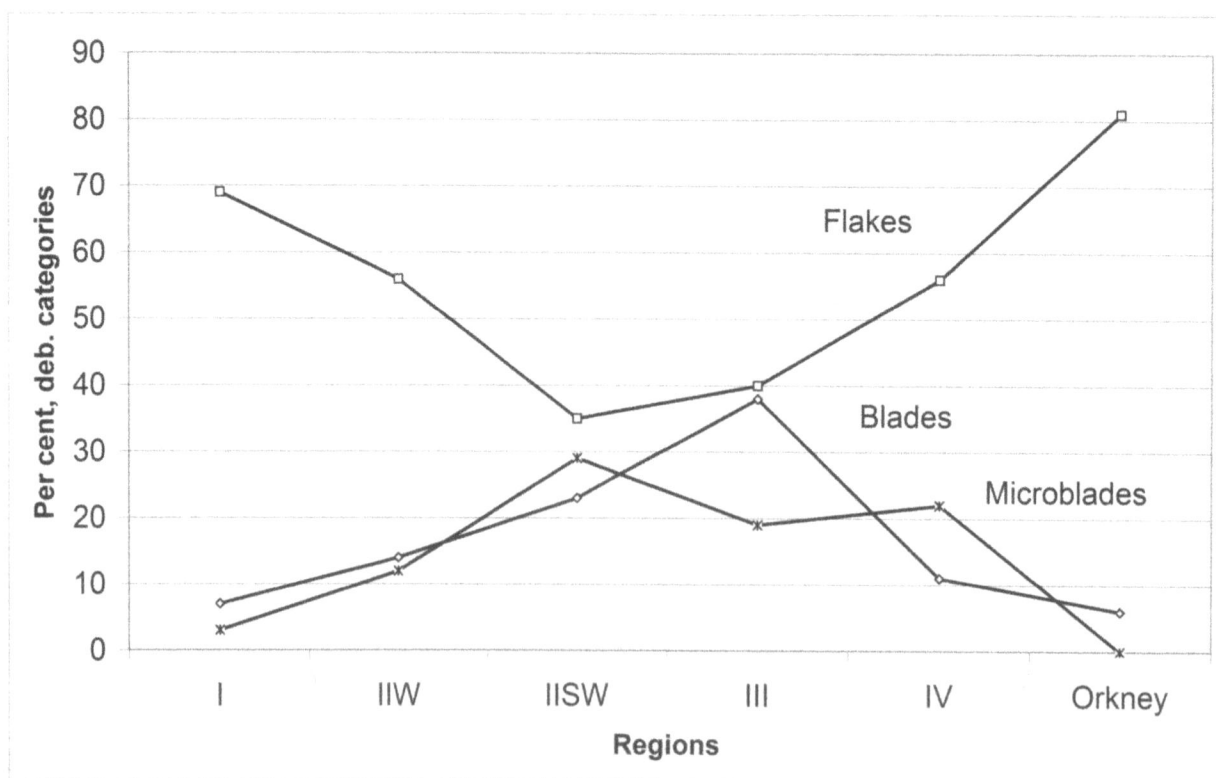

Figure 10. Main debitage categories as a percentage of all debitage, by zone.

One would have expected that the Late Neolithic Blackpark Plantation East collection from Bute (Ballin *et al.* forthcoming) would have included material produced in this manner, but for unknown reasons this was not the case. However, the flint assemblage from the site did.

5.3.3 Raw Material Preferences

The use of different pitchstone forms is quite clearly linked to the production of specific blank types. In periods characterized mainly by microblade production, such as the Late Mesolithic and Early Neolithic periods, practically only aphyric pitchstone was in use (see for example the Luce Bay and Biggar collections), whereas in periods, during which large blades and flakes were produced, lithic reduction usually included some porphyritic material (eg, Barnhouse on Orkney and Blackpark Plantation East on Bute; Ballin forthcoming b; Ballin *et al.* forthcoming).

Due to the link between preferred blank type and raw material preference, the same regional/chronological division displayed in Figure 10 is seen in Figure 11. Figure 11 shows how porphyritic pitchstone is considerably more common in Arran, the west of Scotland and Orkney, than in the south-west, the south-east/east and in the Highland region (for details, see Chapter 4.2.4).

Although the link between technology/raw material preference and chronology seems obvious, it cannot be ruled out that the higher ratio of porphyritic pitchstone in the region immediately north of Arran (Argyll & Bute, including parts of the Southern Hebrides) may, to a degree, be a result of closer territorial links between these areas, either in the sense that they were close allies or that they formed parts of the same social territory. This could possibly have given the area north of Arran access to pitchstone forms (ie, porphyritic material) which were generally not exchanged with the remainder of northern Britain (Chapter 7.5).

5.3.4 Contextual Evidence

Off Arran, worked pitchstone has been recovered from a number of different contexts, which provide dating evidence of varying reliability. Archaeological pitchstone is discussed below in relation to different classes of contexts.

Radiocarbon dated pits

The best dating evidence is that from sealed contexts. Occasionally, pitchstone has been recovered from prehistoric pits, associated with datable charcoal and/or other diagnostic material (archaeological pitchstone from radiocarbon dated contexts is displayed in Figure 12). From a pit near Carzield in Dumfries & Galloway (Maynard 1993), a number of lithic artefacts were retrieved, among which were two small pitchstone bladelets. They were found with sherds of Early Neolithic Carinated Pottery and flakes from a polished stone axehead, the raw material of which macroscopically matches Group VI tuff from Great Langdale in Cumbria. This combination of finds supports the suggestion put forward by Ness & Ward (2001), that in

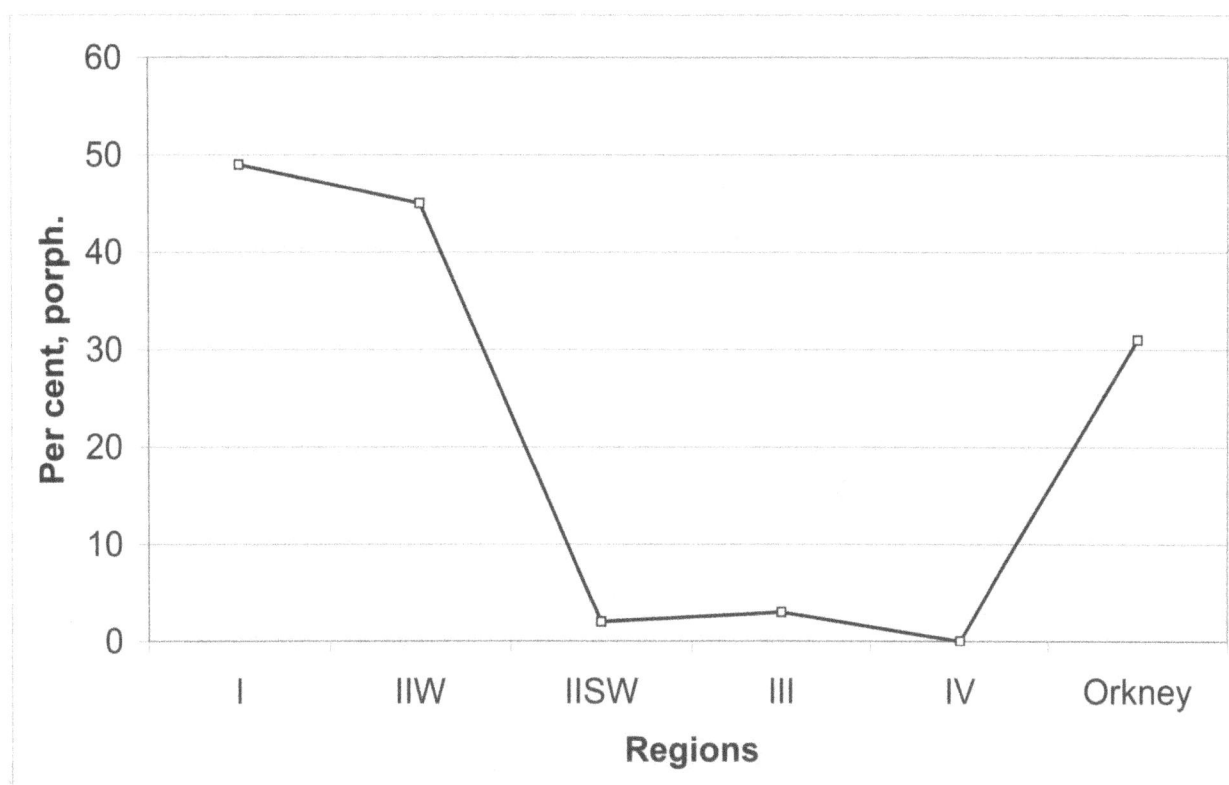

Figure 11. Porphyritic pitchstone as a percentage of all pitchstone, by zone.

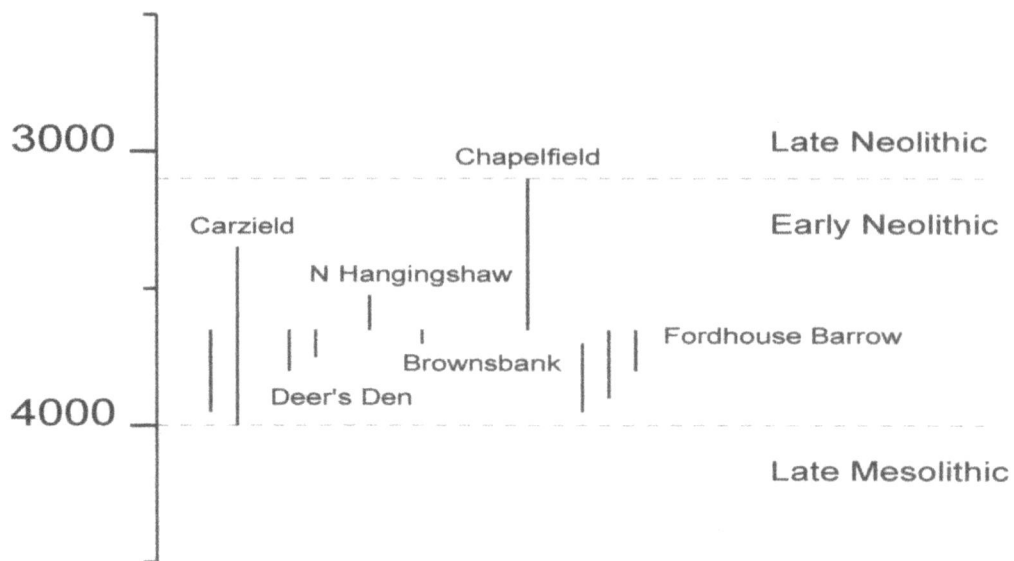

Figure 12. Radiocarbon dated archaeological pitchstone from pits.

the Biggar area pitchstone appears in the Early Neolithic as part of a cultural package including pitchstone, Carinated Pottery and axeheads in Cumbrian tuff. Two charcoal samples from this pit were radiocarbon dated to 3960-3660 cal BC and 4000-3350 cal BC (Beta 68480-1).

Worked pitchstone has also been recovered from two radiocarbon dated pits in the Biggar area in South Lanarkshire. From a pitchstone-bearing pit at Brownsbank, a date was obtained of 3692-3639 cal BC (GU-9303), and a similar pit at Nether Hangingshaw yielded a date of 3640-3520 cal BC (GU-12113) (Tam Ward pers. comm.). The former contained one microblade and a small flake, and the latter a small flake.

A single pitchstone flake was recovered from Pit 1028 at Deer's Den, Kintore, Aberdeenshire (Alexander 2000, 57). It was associated with 61 flint and quartz artefacts, amongst which were a flint leaf-shaped arrowhead. The pit also contained 168 sherds of Early Neolithic Carinated Pottery. Two charcoal samples from Pit 1028 were dated to 3800-3640 cal BC and 3770-3630 cal BC, respectively (OxA-8132-3).

Underneath Fordhouse Barrow, a truncated Early Neolithic long-barrow in Angus (Ballin forthcoming f), 10 pieces of pitchstone were found in a pit. Six of them were unmodified and modified microblades, and nine pieces could be refitted in various groups. The pit also included a bifacial knife in flint. This particular pit was radiocarbon dated to the Late Mesolithic period (4600-4360 cal BC; OxA-8226), but as the pitchstone was associated with a bifacial knife, this date is clearly erroneous. The use of invasive retouch is generally accepted as a post-Mesolithic phenomenon. However, three dates from contexts sealing this pit all yielded dates between 3960 and 3640 cal BC (OxA-8222-4), suggesting

a date for the contents of this pitchstone-bearing pit in the earlier half of the Early Neolithic period.

Nineteen pieces of pitchstone were recovered from three pits at Chapelfield, Stirling, but the associated radiocarbon dates are not very helpful. Two radiocarbon samples from Pit I, in which one piece of pitchstone was found, suggest dates in the eighteenth and nineteenth centuries AD and in the later Mesolithic period (4540-4330 cal BC; OxA-975). The first date is obviously irrelevant, whereas the other is based on a hazelnut shell from the feature's upper fill, which may have entered the pit with its backfill. There is no doubt as to the pit content's general Neolithic date, as the Pit I finds included Neolithic pottery (Squair & Jones 2002, 147).

One radiocarbon sample (Figure 12) dates the finds from Chapelfield Pit VIII, which included much pitchstone (mostly from one core; Donnelly 2002, 172), to 3650-3050 cal BC (GU-7202). This is a very broad and apparently fairly late date. In general, the radiocarbon dates in Figure 12 indicate deposition of pitchstone artefacts in pits in the first half of the Early Neolithic, whereas the date from Pit VIII also covers the later half of the Early Neolithic. Unfortunately, this pit did not contain any diagnostic pottery and should therefore probably be disregarded from this debate. The third pitchstone-bearing pit (Pit IX) was not radiocarbon dated.

Most likely, the two now missing pitchstone 'chips' (in the older archaeological literature 'chip' may refer to any form of flake or blade as long as it is relatively small) from the Roman Fort at Whitemoss, Bishopton, derived from the same form of context as the pits described above. The find was described in the following manner (*Notes on Excavations in the British Isles, 1957*; Proceedings of the Prehistoric Society 1958):

'Beneath the Roman Fort, Professor Piggott found eight shallow oval or circular pits containing black greasy soil, a flint leaf-shaped arrowhead and a scraper, two chips of Arran pitchstone, and pottery of the type found at Bantaskine, Easterton of Roseisle and Lyles Hill, Belfast [that is, most likely Early Neolithic Carinated Pottery; the author's note]'.

Burial monuments

Outside Arran, 53 pitchstone artefacts have been reported from 20 burial monuments. In most cases, the assemblages from burial monuments include one piece only, but the six largest assemblages include from five to 10 pieces (Table 10). Most of the monuments are of Early Neolithic date, and include monuments belonging to Henshall's Clyde group, Bargrennan group, Hebridean Group, Orkney-Cromarty group and Zetland group (although not from Shetland), in long-barrows as well as in round barrows. Another large group is made up of Early Bronze Age burial monuments which will be discussed in greater detail below.

In many of these cases, the finds are clearly residual and embrace artefacts from pre-cairn soils and cairn fills. This category includes the artefacts from Beacharra Cairn (Scott 1964), Cairnderry (Cummings & Fowler 2007), Camster Long (Masters 1997), North Mains (Barclay 1983), Pitnacree (Coles & Simpson 1965), Houston South Mound (Stables 1996) Cloburn Cairn (Lelong & Pollard 1998), and Stoneyburn Farm (Banks 1995). Another large group embraces finds which may have been deliberately deposited in prehistory, either immediately prior to the monument's construction, in connection with the burial, or as part of later ceremonies around the monument. This category includes the artefacts from Fordhouse Barrow (Ballin forthcoming f), Achnacreebeag (Ritchie 1970), Brackley (Scott 1956), Glecknabae Cairn (Bryce 1904), Michael's Grave (Bryce 1904), Monybachach (Scott 1998), Cairnholy (Piggott & Powell 1951), Ord North (Sharples 1981), and Tulach an t'Sionnaich (Corcoran 1966). Apart from Monybachach (see below), which is a cist burial, all other monuments in this category are datable to the Early Neolithic period.

The burials from Kirkburn and Luce Bay have both been dated to the Early Bronze Age, but the status of the pitchstone from these sites is more uncertain. Kirkburn (Cormack 1964) is mainly a Bronze Age cremation cemetery, but the site also includes a number of Early Neolithic pits. Pottery from these pits and cists includes Carinated Pottery, Grooved Ware, Beakers, Food Vessels, and Urns. Pit 34, which contained the pitchstone flake, did not include any pottery or bone, but spatially it is associated with a group of Neolithic pits. The Luce Bay site was, according to a note deposited with the pitchstone find, an Early Bronze Age burial '... with [a] horizontally placed urn'. Nothing is known about the site and the pitchstone artefact, but at Kelvingrove Art Gallery and Museum it is

believed that this may be one of many finds deposited in the early twentieth century by Ludovic Mann (Katinka Stentoft, pers. comm.). However, it is not possible to clarify whether this piece really does represent a Bronze Age deposit, or whether it may be a residual artefact which, for example, entered the burial with the backfill.

As indicated in Table 10, several of the burial monuments yielding pitchstone are datable to the Early Bronze Age and therefore of particular interest to the dating of pitchstone use in general. These monuments include a number of cairns (Stoneyburn), a ring-cairn (Cloburn), a mound (Houston South Mound) and several cists (Kirkburn, Monybachach, Luce Bay), as well as two re-used Neolithic chambers (Brackley and Achnacreebeag). Most of these are discussed above, but the finds from Brackley, Achnacreebeag and Monybachach deserve closer scrutiny.

The Brackley tomb belongs to Henshall's (1972) Clyde group, and in this case, four pieces of pitchstone were recovered near a secondary Early Bronze Age burial inside the chamber. The excavator (Scott 1956, 32) describes a slightly 'messy' chamber, with at least one piece of pitchstone found underneath the Bronze Age paving, and where one Early Neolithic Beacharra B sherd was recovered above the paving. This suggests that the pitchstone could relate to the original Early Neolithic burial, which is supported by the fact that one of the pitchstone artefacts is a true blade.

The situation was much the same in the chamber of the Achnacreebeag passage-grave (Henshall's Hebridean group; 1972). In this case, Early Neolithic pottery was found with, and dominated by, Beaker pottery, and the typo-technological attributes of the flints are consistent with a mainly Early Bronze Age date (Ritchie 1970, 33, 42, 47-50). The chamber also contained jet. Although the conditions in the chamber prevents certain dating of the pitchstone artefacts, the absence of blade material, in conjunction with the fact that two of the three pieces are in porphyritic material (see raw material preference, above), indicate a relatively late date – possibly in the Early Bronze Age.

The Monybachach burial is different. It represents one of three typical Early Bronze Age cists found together at Monybachach in Kintyre, Argyll & Bute, and in the first cist excavated by Scott (1998), five pitchstone flakes were found with an impressive jet necklace, a bronze knife, a whetstone and a piece of worked flint. It has not been possible to inspect the pitchstone finds, as they were not with the jet artefacts in Kelvingrove Art Gallery and Museum. If they were deposited in Campbelltown Museum, they may have been misplaced in connection with refurbishment of that museum, or sometime before. However, the cist represents a primary burial and not secondary interment in, for example, an Early Neolithic chamber or cairn, and as none of the artefacts appear to be blades, there are no reasons why they

CAT	Site	Context
M271	Fordhouse Barrow, Angus	Underneath Early Neolithic long-barrow (Proudfoot pers. comm.).
M013	Achnacreebeag, Argyll & Bute	Possibly from within Chamber 3 (Fig 1 diff to interpret; Ritchie 1970); Henshall's (1972) Hebridean group.
M093	Beacharra Cairn, Argyll & Bute	Probably from pre-cairn soil on SW side of cairn (Scott 1964); Henshall's (1972) Clyde group.
M090	Brackley, Argyll & Bute	A very messy situation in the chamber (EBA re-use of Neo chamber) makes it difficult to determine whether the pitchstone is Neo or EBA, however, 1 true blade suggests pre EBA; Henshall's (1972) Clyde group.
M010	Glecknabae Cairn, Argyll & Bute	Acc T&T (1984) from Chamber 2; acc Henshall (1972) from SE as well as N chambers; Henshall's (1972) Clyde group.
M012	Michael's Grave, Argyll & Bute	From the floor of the chamber; Henshall's (1972) Clyde group.
S072	Monybachach, Argyll & Bute	From EBA cist; associated with jet necklace; unfortunately, the pitchstone artefacts have diappeared.
S013	Cairnderry, Dumfries & Galloway	From pre-cairn soil; Henshall's (1972) Bargrennan group.
M016	Cairnholy I, Dumfries & Galloway	(a) Behind S portal stone in antechamber; (b) from the blocking, ass w carinated pottery; (c) from the clean earth spread in front of N portal; Henshall's (1972) Clyde group.
M066	Kirkburn, Dumfries & Galloway	From Pit 34, one pitchstone flake, no pottery; probably belonging to group of Neo pits next to BA cremation cemetery (Cormack 1964).
M122	Luce Bay, Dumfries & Galloway	It was not possible to identify this burial, but according to the Kelvingrove Museum (Katinka Stentoft pers. comm.), the assemblage was probably recovered and deposited by Ludovic Mann: 'From Bronze Age burial with horizontally placed urn, from Luce Bay'(?).
M320	Camster Long, Highland	Most lithics from pre-cairn soil, a small number from below extra revetments (Masters 1997); Henshall's (1963) Orkney-Cromarty group.
M061	Ord North, Highland	From platform surrounding cairn (Sharples 1981); Henshall's (1963) Orkney-Cromarty group.
M017	Tulach an t'Sionnaich, Highland	3 inches above the floor of the chamber; heel-shaped cairn (Corcoran 1966), Henshall's Zetland group (1963).
M324	North Mains, Perth & Kinross	All pitchstone from fill of barrow, therefore pre-dating the monument (as suggested by the blade/microblade character of some of the pieces); construction radiocarbon dated to 2400-1600 cal BC (GU-1102); non-megalithic round-barrow (Barclay 1983).
M014	Pitnacree, Highland	From old land surface; ring-cairn in round-barrow (Coles & Simpson 1965).
M329	Houston (South Mound), Renfrewshire	From pre-cairn soil, below Bronze Age mound (Stables 1996).
M102	Cloburn Cairn, South Lanarkshire	The pitchstone artefacts were recovered from the barrow fill; Bronze Age ring-cairn (Lelong & Pollard 1998).
M216	Stoneyburn Farm, South Lanarkshire	From topsoil or cairn fill, several from pre-cairn 'black deposit' (incl microblade cores, one barbed-and-tanged arrowhead, and Carinated Pottery); one piece from redeposited turf. Three small Early Bronze Age cairns.

Table 10. Non-Arran pitchstone-bearing burial monuments.

should not be accepted, at the present moment, as the only verifiable pitchstone artefacts datable to the Early Bronze Age period beyond the borders of Arran.

Traditionally (eg, Ritchie 1968; Williams Thorpe & Thorpe 1984), pitchstone artefacts have also been associated with two Bronze Age cemeteries in Fife, Brackmont Mill (four pieces; Ritchie 1968) and Cowdenbeath (one piece; Lacaille 1931). However, in the project's database, both these finds have been characterized as stray finds rather than finds from burials, as the pitchstone artefacts were collected from the surface of the fields in question, rather than in direct association with the Bronze Age burials. Moreover, as two of the four pieces from Brackmont Mill are blades (one unmodified and one modified), they almost certainly pre-date the Bronze Age.

Houses and 'halls'

In the present context, the association of pitchstone artefacts with dwellings is only significant in as far as the building forms are themselves diagnostic. Worked pitchstone has been recovered from several sites on which building remains, or traces of buildings, have been discovered, but usually the pitchstone was not directly associated with the house remains, or the house remains were undiagnostic (eg, Ardnadam, Deer's Den, Chapelfield; Rennie 1984; Alexander 2000; Atkinson 2002).

Basically, only three sites are relevant in this context, namely those of Claish in Stirlingshire (Barclay *et al.* 2002), Warrenfield in Aberdeenshire (Fraser & Murray 2005), and Barnhouse on Orkney (Richards 2005). The houses of the former two sites belong to the group of so-called Early Neolithic 'timber halls', whereas those of the latter belong to the typical Orcadian Skara Brae type houses from the later Neolithic period.

From Claish, two pieces of worked pitchstone were recovered, a flake and a microblade. One was retrieved from a posthole in the east wall (F4), whereas the other was retrieved from an internal posthole (F36). The position of the pitchstone suggests that the pieces are either contemporary with the house, or – if they entered the postholes with the backfill – they may pre-date the building. Neither posthole was radiocarbon dated, but a series of radiocarbon dates from other contexts in the house indicates a date in the range 3970-3520 cal BC (see list of dates in Barclay *et al.* 2002, 99). These dates correspond well with the dates suggested by pitchstone artefacts in pits (Figure 12). At Warrenfield, pitchstone artefacts were recovered from two internal axial pits, which were radiocarbon dated. One chip was found in Pit 30 and two in Pit 50. The former was dated by five samples to 3940-3640 cal BC (SUERC-4038-43), and the latter was dated by a single sample to 3940-3660 cal BC (SUERC-10083). These samples probably date the time when the 'hall' burnt down (Murray & Murray forthcoming). Another three pieces of pitchstone were

retrieved from the pits of structural posts which were not radiocarbon dated. The hall and the pitchstone are also securely associated with Early Neolithic Carinated Pottery (Sheridan 2007, 481).

At Barnhouse, the association of pitchstone artefacts with the Skara Brae type houses is not by traditional 'vertical stratigraphy' (as at Claish and Warrenfield); it is by 'horizontal stratigraphy', where the distribution of artefacts within and outside the buildings suggests a link between pitchstone and houses. One group of artefacts was found around the large hearth of Structure 8 (seven pieces), one outside House 7 (11 pieces), with the remainder having been retrieved from within or immediately outside Houses 10 (two pieces) and 12 (three pieces). The pitchstone group outside House 7, immediately to either side of the entrance (Richards 2005, 45), suggests that this may have been an area where, in good weather, pitchstone was being reduced, or the pieces may have formed part of two so-called 'door dumps' (Binford 1983, 151; also see Ballin forthcoming h). Based on his analysis of the site's radiocarbon dates, Ashmore (2005, 388) suggests that the village may have been constructed around 3100 cal BC, or possibly slightly before, and abandoned about 2900 cal BC, but no later than 2750 cal BC. These dates define Barnhouse as a Late Neolithic settlement, and the Barnhouse pitchstone assemblage is unusual in the sense that it is the only collection outside Arran and western Scotland clearly datable to a period after the Early Neolithic.

Ceremonial sites

This category includes finds from one pit enclosure (Cowie Road), one enclosure/possible henge (Balfarg Riding School), one henge (North Mains), and one setting of standing stones (Duntreath). At Cowie Road in Stirlingshire (Rideout 1997), one flake and one microblade in pitchstone were recovered, both from Enclosure 1. The microblade was found in Pit 34 (Phase 2), whereas the flake was found in Pit 37 (Phase 3). Neither pit has been radiocarbon dated, but a series of five radiocarbon samples from Phase 2 of the enclosure (Rideout 1997, 53) suggests use of these pits in the period 4035-3529 cal BC (AA-20409-13), that is, the first half of the Early Neolithic period. A post hole (PH43) was dated to 3348-3103 cal BC (AA-20414-5), and hazelnut shells from 'fire pit' 59 were dated to the Mesolithic period, 5508-5368 cal BC (AA-20413). Although no pottery was recovered from the two pitchstone-bearing pits in the enclosure, Early Neolithic Carinated Pottery was retrieved from neighbouring pits of Enclosure 1.

The ceremonial site at Balfarg Riding School (Barclay & Russell-White 1993) yielded a fairly substantial assemblage of pitchstone (20 pieces). In the original report, Barclay & Russell-White (ibid., 191) analysed the distribution of the pitchstone, and they noticed the following pattern: where most of the flint is from the enclosure ditch with a substantial supplement in its interior (around Structure 2, the 'timber hall'), most of the pitchstone is from the interior of the enclosure (around the 'timber hall') with only a minor proportion in the ditch. Generally, this suggests that the pitchstone may be contemporary with the Early Neolithic 'hall' rather than with the later enclosure. This is supported by the fact that the assemblage, which is dominated by flakes, includes three unmodified and modified narrow blades. One flake was recovered from posthole F7041 in the 'hall' (Barclay & Russell-White 1993, 85), supporting the proposition that the pitchstone may either be contemporary with this structure or pre-date it.

At the North Mains henge (Barclay 1983), five undiagnostic pieces of pitchstone were retrieved. Unfortunately, all pieces are poorly contexted and it is uncertain whether these pieces are in any way related to the henge. One piece derives from the excavated area outside the ditch; two pieces are from the secondary fills or topsoil of the ditch; and two are from surface collection outwith the site.

In 1972, Euan MacKie carried out excavations at the Duntreath Standing Stones (MacKie 1973). In connection with his investigation, he recovered a small lithic assemblage which included two pieces of pitchstone. It was not possible to find the assemblage in the museum stores, but a photograph of the finds, kindly provided by Dr MacKie, suggests that the pieces are proximal fragments of either blade-like flakes or blades of the sort most commonly associated with earlier Neolithic sites. There had been a fire on the old ground surface, immediately beside the southernmost stone, which left traces of ash and charcoal, and the lithics were found in a context stratigraphically later than this layer. A charcoal sample from the old ground surface was subsequently dated rather broadly to 4500-2500 cal BC (GX-2781), which embraces the period Late Mesolithic to Late Neolithic. As the lithic assemblage from the old ground surface includes not only definitely pre-Bronze Age (and probably pre-Late Neolithic) blade material in pitchstone, as well as the fragment of an Early Bronze Age barbed-and-tanged point, this context is undoubtedly disturbed.

Rock art

Only a small number of rock art sites are known to have yielded pitchstone, namely Ben Lawers in Perthshire & Kinross, and Torbhlaren in the Kilmartin valley, Argyll & Bute. At the former site, a flake and a blade were recovered, whereas the latter site yielded a single chip (Richard Bradley and Andrew Jones pers. comm.). In neither case is the pitchstone (all of which is aphyric) safely associated with the actual rock art site, and in the case of the Ben Lawers pitchstone, the character (ie, width and platform preparation) of the blade would fit better with a date in the middle to later part of the Early Neolithic period than in the Late Neolithic/Early Bronze Age, the period to which rock art sites are traditionally dated.

5.3.5 Association/Disassociation With Other Diagnostic Find Groups

As part of the task of dating pitchstone use outwith Arran, it is relevant to look into which other diagnostic artefact types or materials pitchstone has been safely and routinely associated with, and which artefact types or materials it seems rarely or never to be associated with. The first group includes Cumbrian tuff and Early Neolithic Carinated Pottery, whereas the second group includes Yorkshire flint and Later Neolithic Impressed/Grooved Ware.

The association of Cumbrian tuff with worked pitchstone is mostly intuitive. It was noticed during the many fieldwalking campaigns of the Biggar Museum Archaeology Group (Tam Ward pers. comm.; also Ness & Ward 2001) that the fields and concentrations including multiple finds of pitchstone artefacts also usually yielded Carinated Pottery and Cumbrian tuff (Group VI axeheads and fragments thereof). Although this find combination occurs so frequently in that area that contemporaneity is almost certain, Cumbrian tuff has rarely been found in safe association with worked pitchstone in the Biggar area, or elsewhere.

The only secure find context for this combination is the Early Neolithic pit from Carzield in Dumfries & Galloway (Maynard 1993, 27), which was radiocarbon dated to 3960-3660 cal BC and 4000-3350 cal BC (Beta 68480-1). The total content of the pit was: three flint flakes, two microblades in pitchstone, three flakes struck off an axehead in Cumbrian tuff, and several large sherds of Carinated Pottery. Obviously, there is no doubt of the Early Neolithic date of these Group VI flakes, but unfortunately the abandonment date for Cumbrian tuff use in general is less well dated. Apparently, Cumbrian tuff was exploited through the Neolithic period, and possibly into the Early Bronze Age period (Manby 1979, 72; Smith 1979, 18-19). This means that, without an accompanying radiocarbon date or diagnostic pottery, the presence of Cumbrian tuff in itself would not date the pitchstone artefacts.

Carinated Pottery seems to have been introduced in Scotland at approximately the same time as pitchstone artefacts were beginning to appear in mainland Scotland, and the frequency of finds belonging to the two artefact groups apparently starts to decrease at approximately the same time (probably apart from parts of western Scotland, where pitchstone may have been in use slightly longer). Sheridan (2007) has recently discussed the dating of Carinated Pottery in Scotland, and she suggests the introduction of this type of pottery in Scotland around 4000 cal BC and the continued use of Carinated Pottery and derived forms (eg, Henshall's 'North-Eastern' style) until probably at least 3600 cal BC (ibid., 451-8).

In her paper, Sheridan (ibid., 479) lists Early Neolithic sites where Carinated Pottery has been found. Sites from this list, which also yielded worked pitchstone, are presented in Table 11. Of her 44 sites with this pottery, 25 (or almost 60%) also contained pitchstone.

It is more difficult to deal with negative evidence, that is, material which is rarely or never associated with pitchstone, as 'absence of evidence is not necessarily evidence of absence'. Nevertheless, the trends are pretty clear: On the Scottish mainland, worked pitchstone appears to be rare or absent from contexts with Yorkshire flint or later Neolithic pottery styles.

Yorkshire flint is easy to recognize, partly by its grey to dark-brown/almost black colours, and partly (and more securely) by its soft cortex (when cortex is present). However, in many cases, it is not possible to prove or disprove contemporaneity between the pitchstone and the Yorkshire flint, as sites may have been disturbed. This, for example, is the case at Urquhart Castle at Loch Ness, Highlands, where one piece of pitchstone and five small chips and flakes in Yorkshire flint were found (Ballin 2005b). At this location, the lithics are generally residual

Sheridan no.	Site	County
2	Les Murdie Road	Moray
4	Deer's Den	Aberdeenshire
8	Warrenfield, Crathes	Aberdeenshire
10	Wardend of Durris	Aberdeenshire
11	Dubton Farm	Angus
12	Fordhouse Barrow	Angus
16	Claish	Stirling
17	Cowie Road	Stirling
19	Ratho	Edinburg
24	Wester Yardhouses	South Lanarkshire
25	Weston	South Lanarkshire
27	Brownsbank	South Lanarkshire
28	Melbourne	South Lanarkshire
29	Carwood Farm	South Lanarkshire
30	Biggar Common	South Lanarkshire
31	Nether Hangingshaw	South Lanarkshire
32	Annieston	South Lanarkshire
35	Carzield	Dumfries & Galloway
37	Holywood North	Dumfries & Galloway
39	Pict's Knowe	Dumfries & Galloway
40	Dunragit (Torrs Warren)	Dumfries & Galloway
41	Cairnderry	Dumfries & Galloway
42	Newton, Islay	Argyll & Bute
43	Port Charlotte, Islay	Argyll & Bute
44	Upper Largie	Argyll & Bute

Table 11. Sites from Sheridan's (2007) list of Scottish locations with Early Neolithic Carinated Pottery which also yielded archaeological pitchstone.

as they were found in contexts associated with much later activities relating to the castle.

Although no worked pitchstone has been found securely associated with Yorkshire flint, East Lochside in Angus (Ballin 2005a; Johnson & Ballin 2006) may be the best candidate for a possible association between the two. The site yielded 570 lithic artefacts, mostly in flint, with one piece being in pitchstone and 26 in grey Yorkshire flint. The pitchstone artefact is a small blade fragment, and the width of this piece (10 mm) indicates a date in the middle or later part of the Early Neolithic period. It is in either porphyritic pitchstone or a form of natural glass with relatively large spherulites. Half of the artefacts in Yorkshire flint are debitage, two of which are broad blades (widths 17 and 19 mm), with the other half representing cores and tools. The cores are one discoidal and one bipolar core, whereas the tools include three chisel-shaped arrowheads, five scrapers, and one piece with edge-retouch. Another three chisel-shaped arrowheads may be in local flint.

The lithic chronological indicators, in conjunction with the domination of the site's pottery by Impressed Ware, suggest that the assemblage is almost exclusively later Neolithic (ibid.), although one leaf-shaped point and three pressure-flaked thumbnail-scrapers indicate brief visits to the site in the Early Neolithic and Early Bronze Age periods. Unfortunately, the entire site has been ploughed up, and no artefacts are safely associated with any other. The likelihood of the site's solitary pitchstone blade being contemporary with the exotic flint is supported by the fact that two later Neolithic chisel-shaped arrowheads in pitchstone have been found on the Scottish mainland, namely one in porphyritic pitchstone at Luce Bay, Dumfries & Galloway, and one in aphyric pitchstone at Biggar, South Lanarkshire.

In the Scottish Borders, several sites are known from which large assemblages, almost exclusively in grey Yorkshire flint, have been collected. The most important of these are arguably the assemblages from the neighbouring sites of Overhowden and Airhouse (Sharp 1912; Callander 1928; Stevenson 1948), in the immediate vicinity of the Overhowden henge. These collections include large numbers of chisel-shaped and oblique arrowheads (Table 12) and, when classified and sequenced according to Clark's (1934b) arrowhead typology, the combination of Late Neolithic arrowheads suggests that: 1) both assemblages are later than the material from East Lochside; and 2) the Late Neolithic finds from Overhowden are probably generally later than those from Airhouse.

Three pitchstone blades were recovered from Airhouse – that is, the earlier of the two sites – creating a situation similar to that of East Lochside: some pitchstone may have been exchanged at the time of the earliest part of the Late Neolithic period (at East Lochside associated with Impressed Ware). However, the Airhouse finds are, like those of East Lochside, unstratified finds, and the fact that

Main artefact types	Overhowden	Airhouse	Total	
Oblique arrowheads	17	9	26	
Chisel-shaped arrowheads	17	47*	64	* = incl. triang 'arrowheads' and 'implements'
Leaf-shaped arrowheads	3	19**	22	** = incl. frags, 'spearheads'
Barbed-and-tanged arrowheads	4	9	13	
Scrapers	16	94	110	
Flint-axes		1	1	
Jet		x		

Table 12. Main artefacts from Overhowden and Airhouse (numbers based on Sharp 1912 and Callander 1928).

the Airhouse location also yielded some Early Neolithic objects, makes it impossible to determine safely whether small amounts of pitchstone still trickled through northern Britain at the beginning of the Late Neolithic.

As indicated above by the finds from East Lochside, worked pitchstone may on rare occasions be associated with Impressed Ware, but at the moment pitchstone artefacts have only been found with Grooved Ware at Barnhouse on Orkney (Richards 2005). This may reflect the territorial situation suggested above (also Chapter 7.5), with the west of Scotland being more closely linked to Arran and the source of pitchstone, than the rest of Scotland, and with pitchstone use probably continuing somewhat longer in this area, as it does on Arran itself. Although Orkney is unlikely to have formed part of the territory of, or close alliance between, Arran and the area north of this island, it appears to have been embraced, at least loosely, by the slightly later pitchstone exchange network of western Scotland. In the Biggar area, South Lanarkshire, it has been noted that fieldwalked sites with Late Neolithic Grooved Ware are mostly devoid of pitchstone (Tam Ward, pers. comm.).

At Monybachach on Kintyre, Argyll & Bute (Scott 1998), five flakes of pitchstone were found in an Early Bronze Age cist, with a jet necklace, a bronze knife, a whetstone, and a piece of worked flint. An adjacent cist contained a Food Vessel.

5.3.6 The Dating of Worked Pitchstone Recovered in Areas Beyond Arran – a Brief Summary

Although the territorial aspect and the exchange networks of Early Neolithic northern Britain are discussed at length in later chapters, it is necessary to briefly sketch the perceived territorial structure, as it appears that beyond the island of Arran pitchstone was the subject of exchange for different lengths of time in different parts of the country.

Basically, two different patterns emerged in connection with the above discussion of the chronological evidence: 1) in most of northern Britain, pitchstone use and exchange may have been a largely Early Neolithic phenomenon, whereas 2) immediately north of Arran, in Argyll & Bute and in parts of the Southern Hebrides, pitchstone was

probably in continued use until the end of the Early Bronze Age period, as was the case on Arran itself. On Orkney, pitchstone has been proven to be in use as late as the Late Neolithic period. As microblades are diagnostic of the Early Neolithic period as well as the Late Mesolithic, and as other diagnostic Mesolithic artefact forms are absent on the Scottish mainland, there is at present no evidence to suggest a Mesolithic date for any Scottish pitchstone found outside Arran.

South-west, south-east, east and highland Scotland

In this area, which constitutes the bulk of Scotland, all archaeological evidence implies a probably exclusively Early Neolithic date (with Early Neolithic being defined as post Mesolithic but pre Grooved Ware) for all finds of worked pitchstone:

- Diagnostic types include leaf-shaped and chisel-shaped arrowhead forms;
- The finds are apparently generally products of a microblade/blade industry, with blade widths ranging between *c.* 6 mm and 10 mm, with a small number of blades being up to 12 mm wide;
- The responsible industry aimed at producing these microblades/blades on mainly single-platform cores, with the operational schema characterized by preparation forms such as cresting and platform-edge abrasion – there is no evidence of Levallois-like approaches being applied to reduce pitchstone;
- Aphyric pitchstone dominates markedly, and only two pieces in heavily porphyritic pitchstone have been recovered (one from the Luce Bay area and one from the Biggar area);
- Pitchstone from radiocarbon dated pits is generally chronologically restricted to the earlier half of the Early Neolithic period, as is pitchstone from the so-called 'timber halls';
- Finds from burial contexts seem to be exclusively Early Neolithic, as it was possible to generally reject a series of potential associations between pitchstone artefacts and Early Bronze Age graves and cemeteries;
- It was not possible to safely link any archaeological pitchstone to Late Neolithic ceremonial complexes, such as enclosures/henges, standing stones or rock

art sites. Pitchstone found at these sites is most likely residual. The only ceremonial site which could be safely associated with worked pitchstone was the pit enclosure of Cowie Road, Stirlingshire, which was dated to the first half of the Early Neolithic period.

Although the finds from Ballygalley in Northern Ireland have not yet been published, preliminary notes on the site's pitchstone assemblage (eg, Simpson & Meighan 1999; Preston *et al.* 2002) indicate that, in this region, pitchstone use was also an Early Neolithic phenomenon.

The area immediately north of Arran

This area is characterized by the same range of Early Neolithic finds in pitchstone as the region described above, but some finds from western Scotland are apparently somewhat later:

- Pitchstone assemblages from the western part of Scotland frequently include broader blades than assemblages from the south-west, south-east, east and highland regions (at Blackpark Plantation East on Bute up to 30 mm wide);

- No pitchstone assemblages were manufactured in Levallois-like technique (although some were associated with flint assemblages produced in this way). Instead, the crudely porphyritic material was produced on plain single-platform cores with a minimum of core preparation, unlike the typical Early Neolithic pitchstone cores with their carefully abraded platform-edges;
- The assemblages generally include more porphyritic material, mostly in the form of lightly porphyritic pitchstone, but some assemblages include or are dominated by heavily porphyritic glass, such as many (if not most) assemblages on Bute;
- This is the only part of Scotland, where the association of pitchstone artefacts with Early Bronze Age burials has been deemed either 'probable' (eg, the re-used Early Neolithic chamber at Achnacreebeag) or 'almost certain' (eg, the cist burial at Monybachach).

At present, Orkney forms a small exceptional enclave, somewhat outside the Arran/West of Scotland 'territory', but with pitchstone deriving from certain Late Neolithic contexts.

6. PITCHSTONE TECHNOLOGY

6.1 Introduction

Pitchstone industries are, like industries in other raw materials, mainly defined by a combination of the availability, properties and limitations of the raw material, and the general lithic approaches of the period(s) embraced by those industries. Below, an account is given of pitchstone availability and properties, as well as of the main operational schemas experienced in connection with Scottish assemblages of archaeological pitchstone.

6.2 Pitchstone Procurement – the Technological Aspects

As stated by Williams Thorpe & Thorpe (1984), most (if not all) archaeological pitchstone was obtained from the Isle of Arran in the Firth of Clyde (Chapter 3). The socio-economical mechanisms behind the pitchstone exchange are discussed in Chapter 7.

The character of pitchstone artefacts from different periods suggests that it may be possible, in procurement terms, to sub-divide the pitchstone-using era into two stages, namely the Mesolithic period and the post-Mesolithic period. The analysis of assemblages from the Mesolithic, such as Auchareoch in southern Arran (Affleck *et al.* 1988, 46), indicates that initially pebble sources were an important element of pitchstone procurement, whereas the procurement of pitchstone in later periods seems to have focused predominantly on primary sources.

This development is probably due to the fact that the first people to settle on Arran favoured flint, which they knew from their place of origin in other parts of the British Isles. Pitchstone was a new material and was procured as a supplement to flint, and as such the island's relatively sparse pebble sources were sufficiently rich to serve the needs of this pioneer phase. However, over time, the use of pitchstone became more common and in the latest part of the Mesolithic period, as well as in the following periods, volcanic glass clearly dominates Arran's lithic assemblages. With the onset of the Early Neolithic exchange in pitchstone, between Arran and other parts of northern Britain, the need for more abundant sources grew, and Arran's inhabitants shifted their focus from pebble sources to the island's many, and substantial, dykes and sills (Ballin & Faithfull forthcoming).

As Scotland's Later Mesolithic/Early Neolithic industries were orientated towards the manufacture of microblades, flawless pitchstone was required. Consequently, the first quarrying of pitchstone must have taken place in eastern Arran, where most pitchstone is aphyric and has excellent flaking properties, although bands of aphyric material in sheets of porphyritic pitchstone could also have been exploited (Ballin & Faithfull forthcoming). Later, when broader and thicker blades and flakes were required and produced, it became possible to also exploit outcrops in northern, western and southern Arran, where porphyritic sources dominate.

In connection with the author's survey of pitchstone sources on Arran (Ballin & Faithfull forthcoming), many exposed pitchstone surfaces were inspected for signs of prehistoric quarrying (cf. Ballin 2004b), but none was seen. This is probably the result of two factors: firstly, if shore dykes were exploited, any signs of quarrying would have been destroyed by coastal erosion, and secondly, if inland dykes and sills were exploited, any worked surfaces would have been destroyed by weathering. Pitchstone is a relatively soft and brittle material, and in many cases it is characterized by the presence of closely spaced planes-of-weakness. In prehistory, the Great Sill at Dun Fionn must have been the source of much of the pitchstone used on Arran or for 'exportation' to the mainland, but it appears somewhat volatile, and at times blocks the size of houses fall down onto the slope leading to the shore.

Most probably, it will never be possible to prove prehistoric pitchstone quarrying on Arran by the examination of quarried rock-faces, but indirect evidence supports the notion of quarrying operations on the island. Most raw pitchstone found in connection with pitchstone-bearing sites on the mainland is in the form of tabular scrap (as for example the large assemblage of burnt tabular material recovered at Torrs Warren in Dumfries & Galloway; Cowie 1996), rather than collected beach pebbles (one exception being the pitchstone pebble from Achnahaird Sands in the Scottish north-west; Ballin 2002b). Tabular scrap can only have been obtained from primary sources.

6.3 Pitchstone Flaking Properties

The choice of operational schemas, as well as the specific shape of many artefact types, is the direct result of the attributes and flaking properties of the exploited raw material. In terms of pitchstone, the most important, technologically relevant raw material attributes are:

- The raw material's tabular character;
- Nodule size;
- The presence/absence of crystalline inclusions;
- Hardness/brittleness;
- A tendency of blanks to curve along the flaking axis.

6.3.1 Tabular Character

The raw material's tabular character meant that, mostly, there was no need to decorticate nodules, as no traditional cortex was present. Occasionally, edges of tabular pieces may have served as preformed crests, or guide ridges, but it appears that, in many cases, tabular pieces were sufficiently irregular to require core preparation, explaining the relatively large number of crested pieces in pitchstone assemblages. However, many of these crested pieces may be secondary specimens, that is, not from the initial core preparation, but from later adjustment of the cores, for example between blade series. In other cases, the surfaces of the planes-of-weakness, which determined the tabular shape, were used as preformed platforms.

There are many examples of how the tabular shape of the raw material determined the specific design of artefact types, and some artefact forms unique to pitchstone assemblages are based on the material's tabular point-of-departure. A common type in pitchstone collections, for example in the Biggar area in South Lanarkshire (Ballin & Ward 2008), is the single-platform microblade core with a flat 'back-side' (the face opposite the flaking-front). In contemporary flint collections, these cores would probably have been conical or bullet-shaped. At Sliddery Farm on Arran, 22 small almost identical discoidal scrapers and end-scrapers were collected; half of these are on flakes and the other half on tabular pieces with flat 'under-sides'. A core type unique to pitchstone assemblages is the flat tabular piece where small blades were detached from one

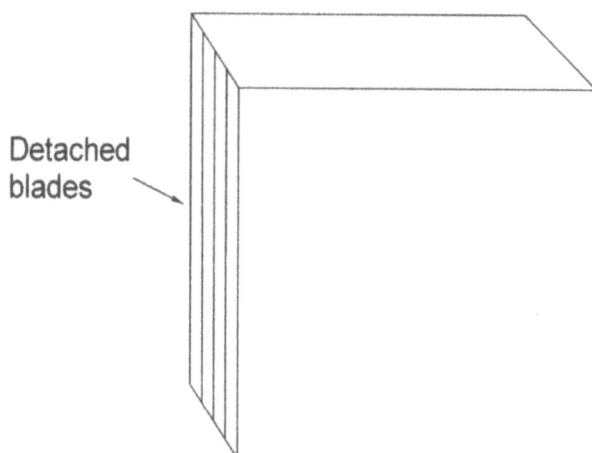

Figure 13. A tabular core.

narrow end in a motion parallel to the planes-of-weakness, rather than perpendicular to those (Figure 13). Basically, this is the same principle experienced in connection with Scandinavian handle-cores.

6.3.2 Nodule Size

Many aphyric pitchstone forms tend to have closely spaced planes-of-weaknesses (Ballin & Faithfull forthcoming), and as a result they break into rather small tabular pieces. Consequently, cores as well as tools based on this type of raw material must be small. This was no problem in the Late Mesolithic period, or in the earliest part of the Early Neolithic period, where lithic production focused on the manufacture of diminutive microblades and tools of small size. However, in the later part of the Early Neolithic period, as well as in the Late Neolithic period, where the industries focused on the manufacture of larger blades, it was necessary to source pitchstone differently. As many porphyritic pitchstone forms have less closely spaced planes-of-weakness, such as those along Arran's west-coast (Ballin & Faithful forthcoming), pitchstone with crystalline inclusions now became relevant.

The size of the available nodules may also be the reason why core rejuvenation flakes, or core tablets, are so relatively rare. Due to the small size of the tabular scrap, the microblade and blade cores were small and, subsequently, they were exhausted relatively quickly. As a consequence, cores were frequently discarded after only one or two blade series and often rejuvenation of the cores' platforms was not needed.

6.3.3 The Presence/Absence of Crystalline Inclusions

As explained in Chapter 3, the many known pitchstone sources on Arran can be subdivided into two main forms, namely aphyric and porphyritic material, or material without and with crystalline inclusions. Aphyric material is generally found in eastern Arran or as bands in porphyritic pitchstone, whereas porphyritic pitchstone dominates the forms found in northern, western, and southern Arran.

Examination of pitchstone assemblages from Scottish museums indicates that, in the Late Mesolithic as well as at the very beginning of the Early Neolithic period, pitchstone assemblages were dominated by microblades (eg, the bulk of the material from Auchategan, Biggar, and Glen Luce; Ballin 2006a; Ballin & Ward 2008; Williams Thorpe & Thorpe 1984), whereas the blades gradually grew broader and thicker through the period and, at the end of the Neolithic period, assemblages were dominated by broader blades (eg, Machrie Moor on Arran, Barnhouse on Orkney, and Blackpark Plantation East on Bute; Haggarty 1991; Ballin forthcoming b; Ballin *et al.* forthcoming).

The examined pitchstone collections also suggest that, outside Arran, the use of porphyritic pitchstone increased through the Neolithic period. In terms of functionality, it would simply not have been possible to produce microblades or narrow broad blades (with microblades and broad blades being narrower and broader than 8 mm, respectively; eg, Wickham-Jones 1990, 73; Ballin 1996a, 9) in the porphyritic pitchstone encountered at, for example, Blackpark Plantation East. A devastatingly large proportion of the blades would simply have broken, or suffered platform collapse, as a consequence of the raw material's huge crystalline inclusions. Microblades in pitchstone must be based on aphyric material.

However, this is a typical chicken-and-egg situation: what came first? Did people start producing larger blades and, consequently, required more pitchstone, as well as pitchstone in larger nodules (see Chapter 6.3.2, above); or did people start using porphyritic pitchstone for other reasons, which allowed them to expand their blade production and produce larger blanks? As the trend towards the manufacture of larger blades is mirrored throughout Britain, no matter which raw materials were exploited in the various regions, the procurement of porphyritic pitchstone must be the secondary element of this equation, that is, a welcome answer to a need for more pitchstone in larger nodules.

6.3.4 Hardness/Brittleness

Compared for instance to flint, pitchstone has advantages, as well as weaknesses. As obsidians and pitchstones are forms of glass, they flake well when they are without larger crystalline inclusions, and it is possible to produce cutting-edges in this material which are much sharper than similar edges in flint. As stated by Whittaker (1994, 19), natural glasses will produce edges that are only a few molecules thick, and obsidian/pitchstone knives are, for example, better surgical implements than similar pieces in steel, as they produce precision cuts which heal with little scarring.

However, pitchstone is also less hard and more brittle, which limits its use somewhat. On Moh's exponential scale from 1 to 10, flint has a hardness of 7 whereas pitchstone has a hardness of *c*. 5-5.5. This means that the working-edges of pitchstone implements dull much faster than those of flint tools and, subsequently, they will either have to be abandoned or re-sharpened. This softness is also the reason why much pitchstone from plough soil has scratched surfaces after having rubbed against harder materials such as flint, quartz, and modern agricultural machinery. A large proportion of the finds from Blackpark Plantation East on Bute (Ballin *et al.* forthcoming) was heavily abraded from movement in the plough soil, and in several cases it was difficult to recognize these more rounded pieces as artefacts, and not pebbles. They were also heavily scratched.

The brittleness of pitchstone means that it splinters easily and, as a consequence, gentler technological approaches (soft percussion and pressure-flaking) seem to give the best results in terms of lithic production. Although the material responds well to hard percussion, some reduction techniques are evidently too violent to assure formal control of the end-products, such as the hammer-and-anvil technique. Bipolar technique was clearly applied occasionally, as evidenced by the recovery of small numbers of bipolar cores, but it seems to have been used sparingly and probably mainly as an *ad hoc* way of dealing with acute problems, rather than as part of a systematic operational schema. The use of rhyolite in Early Neolithic SW Norway (the dominating material of that region in that period) is characterized by the same avoidance of hammer-and-anvil technique, with a similarly low proportion of bipolar cores. Most likely, this was also a result of material brittleness (Ballin 1999, 365).

Although pitchstone may tend to produce sharper edges than flint, the fact that it is relatively soft and brittle has resulted in the total absence of serrated pieces in this raw material. Some Neolithic sites are characterized by large numbers of serrated pieces in flint (eg, Saville 2002; 2006), but the modification of an exceptionally thin and sharp pitchstone cutting-edge by the removal of multiple, closely spaced micro-chips obviously weakens the edge so much that the production of this tool type in pitchstone was completely abandoned. It is probably an additional factor that some serrated pieces were intended to function as saws for the splitting of relatively hard materials, such as bone, antler and wood, or to harvest/process tough plant materials, such as twigs, grasses, grain, and reeds (Juel Jensen 1994, 62).

6.3.5 Curvature

Generally, blanks in pitchstone have an exaggerated tendency to curve along the long axis (Plate 28), frequently causing blades to overshoot and remove the apex of the cores. This results in many cores having distinctly curved surviving apexes. Occasionally, a curving apex was used as a second platform (Figure 14).

Figure 14. An S-shaped opposed-platform core, where the curving apexes (produced by overshot blades) were used as platforms; from Cornhill Farm in the Biggar area, South Lanarkshire (drawn by Mrs Sandra Kelly).

Figure 15. A Glen Luce Type discoidal core from the Biggar Gap Project, South Lanarkshire (drawn by Mrs Sandra Kelly).

The examination of the rich pitchstone finds from the Glen Luce area in Dumfries & Galloway allowed the definition of a specific form of discoidal core, which is rarely (if ever) seen in other raw materials. In a sense, this type is a hybrid core form, with elements from discoidal cores and cores with two platforms at an angle. It is most certainly discoidal, in terms of its general shape, but the microblades detached from the two faces are orientated at perpendicular angles to each other (Figure 15; the principle is explained in Figure 9). In contrast to this, most typical cores with two platforms at an angle are rather cubic specimens. It is possible that the creation of this core type is also a result of the pitchstone blades' exaggerated tendency to curve along the long axis. The discoidal cores from the Biggar area all belong to this core type (Ballin & Ward 2008, 11).

6.4 Operational Schemas

Today, it has become common practice to characterize lithic industries in terms of their operational schemas ('*chaîne opératoire*'; Leroi-Gourhan 1965; Lemonnier 1976). In a paper on the chert assemblage from Glentaggart in South Lanarkshire (Ballin & Johnson 2005, 77), the author defined a basic methodological approach, which will also be followed below. This approach focuses on a number of key steps in the reduction process, such as procurement, core preparation and core rejuvenation, successive steps of blank production, and final abandonment of the process and rejection of the exhausted core and other waste products (ibid., Figures 13 and 14). Pitchstone procurement was dealt with above.

In the following chapters, two operational schemas for pitchstone blank production are characterized on the basis of a number of examined assemblages. These are the schemas of the Early and Late Neolithic periods, which are both highly distinctive, as well as diagnostic. Little is known about the Mesolithic pitchstone industries, as none has been characterized in detail in the archaeological literature, and none was available for closer scrutiny during the SAPP. However, towards the end of the present project, Glasgow University Archaeological Research Division (GUARD) kindly allowed access to the rich lithic assemblages from their recent excavations on Arran (Donnelly & Finlay forthcoming), and it was clear that large proportions of this material was datable to the later part of the Mesolithic. The pitchstone microblades from this period appear to have been manufactured very much in the same manner as the microblades and narrow blades of the following Early Neolithic period, and the Early Neolithic operational schema presented in the following chapter most probably also describes the technological approach of the later Mesolithic. This is supported by examined Mesolithic assemblages in other materials, such as flint and chert (Ballin & Johnson 2005; Ballin forthcoming e).

The operational schemas of the Bronze Age pitchstone industries are not dealt with below, as no chronologically unmixed and numerically acceptable (that is, representative) Bronze Age assemblages are available. However, these industries are probably ordinary flake industries, as in Britain blade production ceased before the end of the Late Neolithic period (see discussion in Ballin forthcoming a; g). Based on contemporary collections in other lithic raw materials (for example quartz assemblages from Shetland and the Western Isles; cf. Ballin forthcoming j), Early Bronze Age pitchstone assemblages most likely follow a relatively systematic, although basic, operational schema (probably: single-platform core ⇒ core with two platforms at an angle ⇒ irregular core), whereas Later Bronze Age industries may follow more unschematic approaches (cf. Ballin 2002a).

The two main technological approaches are characterised on the basis of case studies, with the material from Biggar in South Lanarkshire and Auchategan in Argyll & Bute (Ballin 2006a; Ballin & Ward 2008) representing the Early Neolithic operational schema. The Late Neolithic operational schema is represented by the finds from Barnhouse on Orkney, Machrie Moor I/XI, and Blackpark Plantation East on Bute (Haggarty 1991; Ballin forthcoming b; Ballin *et al.* forthcoming).

6.4.1 The Early Neolithic Approach

In Scotland, the Early Neolithic period is characterized by the production of microblades and narrow blades in aphyric pitchstone. True microblades (that is, blades narrower than 8 mm) are typical of the earliest part of the period, but these gradually develop into wider and wider blades through the remainder of the period (also see Chapter 5). Pitchstone blades from the many Biggar assemblages (Ballin & Ward 2008) generally have widths immediately to either side of the blade/microblade cut-off measure of 8 mm (the composition of the debitage from the large Biggar collection is shown in Table 13). Pitchstone blades from the Auchategan assemblage (Ballin 2006a), which is thought to represent two visits to the site during the Early Neolithic, have average dimensions of 31 x 13 x 8 mm, whereas the microblades are in the order of 23 x 7 x 5 mm.

	Number	Per cent
Chips	61	10.6
Flakes	247	42.9
Blades	124	21.5
Microblades	127	22
Indeterminate pieces	4	0.7
Crested pieces	12	2.1
Platform rejuvenation flakes	1	0.2
TOTAL	576	100

Table 13. The Biggar pitchstone collection. Debitage and preparation flakes.

The material from Biggar, as well as that from Auchategan, includes preparation flakes. Most of these are crested pieces, with core preparation flakes being less numerous. At Biggar, the crested piece to core tablet ratio is 12:1, whereas at Auchategan it is 3:1. Being the larger collection, the ratio of the Biggar material is probably more statistically likely to represent Early Neolithic pitchstone technology in general. As mentioned above, one reason for the relative rarity of core tablets is probably the generally small size of the tabular scrap on which mainland pitchstone assemblages appear to have been based. Due to this fact, cores were very small from the onset of the production process, and each core would probably yield a relatively small number of successive blade series, making core rejuvenation less necessary.

Prior to commencement of blank production, and between the individual blank series, the platform-edges were carefully trimmed and subsequently abraded. This provided the platforms with a more rounded edge, which was stronger than an untreated edge, and platform collapse was generally prevented. Careful preparation of the cores' platform-edges may have been particularly pertinent in connection with the production of microblades and narrow blades in pitchstone, as this raw material is more brittle than most other common lithic raw materials in Scotland. The platforms themselves are mostly plain and unprepared. The microblades and blades were generally detached by the application of soft percussion.

The composition of the Biggar collection's cores (Table 14) give an impression of some of the choices made in connection with Early Neolithic pitchstone reduction. The production of microblades and narrow blades was based mainly on small single-platform cores, which would be transformed into cores of lower rank (that is, cores with more platforms, some of which would be 'dead', or abandoned, platforms), as part of the ongoing management and rejuvenation of the cores' shape. Given that enough mass was left in the cores, single-platform specimens would be transformed, preferentially, into opposed-platform cores but, if the shape of individual, spent single-platform cores was chunky and irregular, they would be transformed into cores with two platforms at an angle. Later, dual-platform cores might be transformed into irregular cores, that is, cores with three or more platforms (also referred to as multi-platform or multi-directional cores). The composition

	Numbers	Per cent
Single-platform cores	18	36.7
Opposed-platform cores	13	26.5
Cores w two platf at an angle	5	10.2
Discoidal cores	6	12.3
Irregular cores	7	14.3
TOTAL	49	100

Table 14. The Biggar pitchstone collection. Cores.

of the Biggar cores is mirrored by the smaller assemblage from Auchategan (Ballin 2006a, 17). Interestingly, the large Biggar collection includes no bipolar cores in pitchstone, which may partly be due to the brittle character of volcanic glass (Chapter 6.3.4), but it may also be due to the fact that, at the end of the above transformation sequence, any core remains would be exceedingly small and useless.

The small discoidal pitchstone cores, which in the Early Neolithic period mostly belong to the Glen Luce Type (Chapter 6.3.5), are difficult to place in the operational sequence of single-, dual- and multi-platform cores. They appear to follow a relatively strict mental template, where microblades were manufactured on thin tabular pieces, and where the production is based on the simultaneous microblade production from two parallel flaking-fronts, but where blanks are detached from perpendicular directions. Given the pitchstone blanks' tendency to curve strongly along the long axis, the resulting core would automatically acquire the regular discoidal shape shown in Figure 15. This core type seems to represent deliberate planning and is probably not part of the above transformation sequence. Once the discoidal shape had been acquired, it would be very difficult to transform this core type into any other type of core by adding new platforms, and only bipolar reduction would have been a logical option, that is, if this approach had been suitable for pitchstone reduction. It should be tested, as future pitchstone assemblages are recovered and dated precisely, whether discoidal cores of Glen Luce Type are diagnostic of a particular Early Neolithic phase.

During the Early Neolithic period, the same general range of tools was produced as in other lithic raw materials. Only serrated pieces are absent from Early Neolithic pitchstone assemblages, probably due to the combination of raw material properties (brittleness) and the use of serrated pieces in the processing of relatively hard materials (discussed in Chapter 6.3.4). The general composition of pitchstone assemblages differs somewhat from that of for example flint assemblages, where simple edge-retouched pieces are more common in pitchstone, if not dominating. This may partly be due to the specific uses to which pitchstone tools were put, but to prove this point 'beyond reasonable doubt' future use-wear analysis of pitchstone assemblages is needed.

Invasively retouched pieces are known in pitchstone, but although pitchstone, being a form of volcanic glass, should be well suited for the reduction by invasive retouch (Inizan *et al.* 1992, 17), such pieces are relatively rare in this material. Basically, the only forms of invasively retouched implements known from Early Neolithic pitchstone assemblages are leaf-shaped arrowheads and plain flake knives with rather expedient scale-flaking along one edge.

6.4.2 The Late Neolithic Approach

The fact that the three key assemblages (Barnhouse, Machrie

45

Moor I/XI and Blackpark Plantation East; Haggarty 1991; Ballin forthcoming b; Ballin *et al.* forthcoming), despite obvious similarities, differ somewhat technologically, means that it is more difficult to produce a general picture of Late Neolithic technological approaches. At present, it is impossible to determine whether these differences are due to chronological or regional differences. Two of the examined cases (Machrie Moor I/XI, Blackpark Plantation East) are from the Arran/Argyll & Bute area and one (Barnhouse) is from Orkney. Examples of Late Neolithic technological approaches have not been identified outside these areas (Chapters 5 and 7).

Although the use of aphyric pitchstone continues into the Late Neolithic period, assemblages now seem to include increasingly large amounts of porphyritic material. This trend may reflect the fact that blades grow broader and thicker through the Neolithic period, with broad, thick blades not being as demanding, in terms of raw material purity, as microblades (Chapter 6.3.3; also see Table 15). The three examined assemblages all represent blade industries, although the three sets of blades differ considerably. The blades from Barnhouse, which are based on lightly porphyritic pitchstone, are relatively narrow, with widths up to *c.* 15 mm. The blanks from Machrie Moor, which are generally in coarsely porphyritic pitchstone, were not measured, but the general impression is that the Late Neolithic blades from the two Arran sites are as wide as, if not slightly wider than, the blades from Barnhouse. At Blackpark Plantation East, a small number of blades in aphyric pitchstone are of roughly the same sizes as the Barnhouse pieces, whereas several blades in coarsely porphyritic pitchstone have widths between 20 and 30 mm.

The pitchstone assemblages from Barnhouse and Machrie Moor are both characterized by the application of the Levallois-like technique. This approach has been shown, via analysis of Scottish Late Neolithic flint assemblages, to represent the period's main reduction technique. As described in Ballin (forthcoming a), the *Levallois-like* technique is quite similar to the *Levallois* technique (Roe 1981, Fig. 3:9) which was in use during the Lower and Middle Palaeolithic periods, although it is possible to define subtle differences between the Levallois and the Levallois-like techniques.

The point-of-departure of Levallois-like reduction is the preparation of a so-called 'tortoise' core rough-out (Figure 16), with a domed, in most cases corticated 'under-side', and a slightly less domed flaking-front (I-II). At one end, a

platform is prepared by fine faceting of the platform surface (III). In contrast to the way Palaeolithic Levallois cores were prepared, the preparation of Levallois-like cores includes the formation of two parallel crests, or guide-ridges, along the two lateral sides of the flaking front. It is thought (Ballin forthcoming a) that the main purpose of reintroducing this Palaeolithic core form in the British Late Neolithic period was the wish to be able to produce broad specialized flakes for chisel-shaped arrowheads, and long slender blades for serrated pieces and other cutting implements, from the same parent pieces. Broad flakes would be produced from the centre of the core's flaking-front (IV), whereas slender blades were produced from the area under, and around, the two highly regular crests (Figure 17).

In many cases, the final waste product of this process was a core which is clearly recognisable as a spent Levallois-like core but, as described in connection with the analysis of the finds from the Stoneyhill site in Aberdeenshire (Ballin forthcoming a), many exhausted cores lost most of their original attributes during reduction, leaving a so-called 'flat core' without the typical tortoise flaking-front of the early-stage Levallois-like cores, and without the typical lateral sides. Only the slightly domed, untouched 'under-side' and remains of the finely faceted platform survive. In connection with the excavations at Late Neolithic Barnhouse on Orkney (Richards 2005), a very small, totally exhausted pitchstone-core was recovered and, surprisingly, all its Levallois-like features are still intact (Plates 29-30).

The pitchstone assemblages from Barnhouse and Machrie Moor include several of the typical waste products of Levallois-like production, such as discarded blanks with finely faceted platform remnants, crested pieces and the Levallois-like cores themselves. In both cases, parallel production of blanks was clearly being carried out on traditional single-platform cores. The preferred way of detaching flakes and blades was now hard percussion. As shown in connection with the analysis of the Stoneyhill material (Ballin forthcoming a), trimming had now become rarer, but abrasion of platform-edges was still being carried out, frequently in conjunction with the typical Late Neolithic fine faceting of the platforms. Abraded platform-edges are also present in the Barnhouse and Machrie Moor collections.

Apparently, the Levallois-like technique was not applied at Blackpark Plantation East (Ballin *et al.* forthcoming) as part of the pitchstone reduction, and blanks seem to have been detached from traditional single-platform or simpler

	Barnhouse	Machrie Moor	Blackpark Plantation East
Blade width	Rel. narrow	Broad	Very broad
Porphyridity	Finely porphyritic	Coarsely porphyritic	
Levall. approach : ordinary platform	Levallois-like		Ordinary platform

Table 15. Late Neolithic pitchstone assemblages. Basic technological trends.

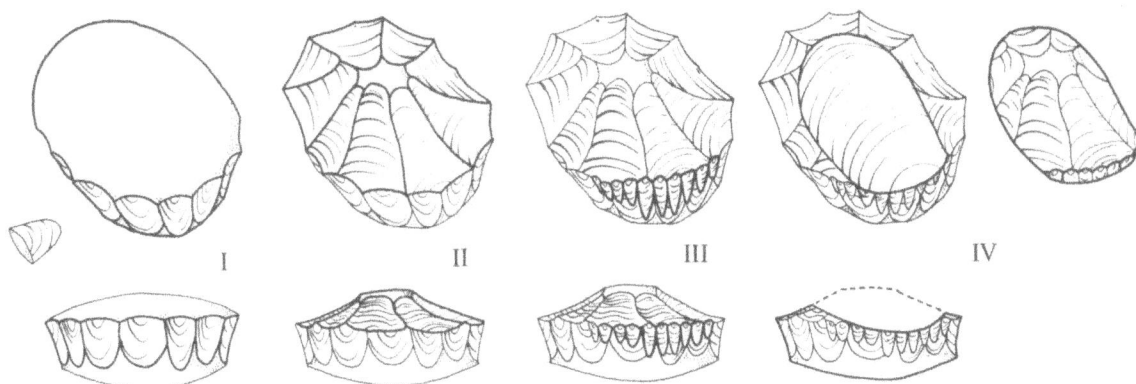

Figure 16. The operational schema of the Late Acheulean / Mousterian Levalloisian (Roe 1981, Fig. 3:9): I. Basic shaping of nodule; II preparation of domed dorsal surface; III. preparation of faceted striking platform on core; IV. the flake and the struck core, with their characteristic features. Drawn by the late M.H.R. Cook.

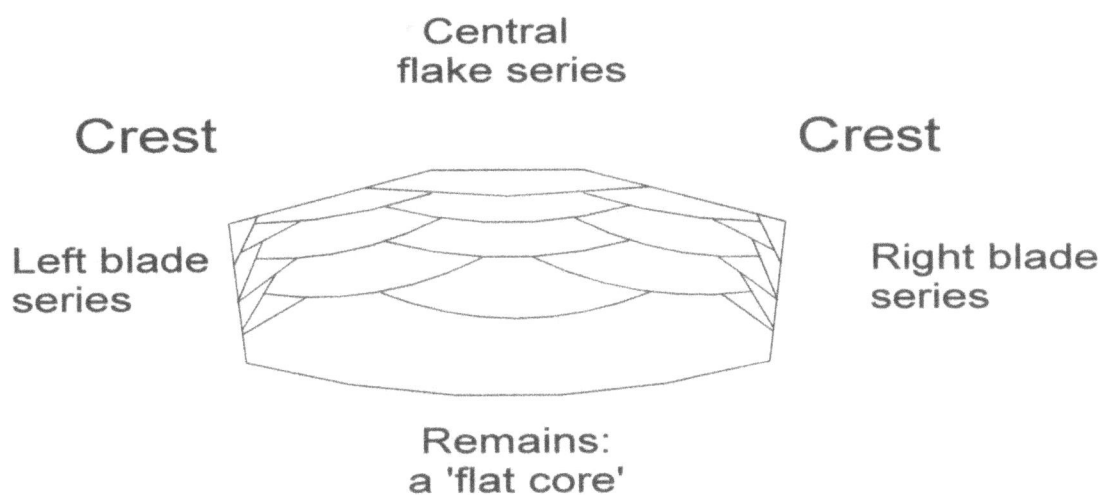

Figure 17. The Levallois-like reduction sequence: first, a number of crested blades and ordinary blades are detached from the core's flanks near the two lateral guide ridges; then, a series of flakes are detached from the relatively flat central flaking front.

core forms. Most pitchstone blades from this site have untrimmed plain platforms, but a number of surviving single-platform cores has trimmed platform-edges. However, finely faceted platform remnants were noticed amongst the small number of flint flakes accompanying this large pitchstone assemblage. At present, it is not possible to explain this difference between the Bute assemblage and the other two assemblages but, most likely, slight differences in the raw material's porphyridity, that is, the size and number of crystalline inclusions, play a part.

The porphyritic pitchstone from Barnhouse includes only few and small inclusions; the porphyritic pitchstone from Machrie Moor includes more frequent and considerably larger inclusions; and – although this is presently a subjective impression – the pitchstone from Blackpark Plantation East is probably even richer in large phenocrysts. The phenocrysts were clearly a problem to people at the Bute site, as they caused many platforms to collapse during detachment, many blanks snapped through phenocryst-rich zones during production and use, and many edges were

damaged where phenocrysts fell out of their pitchstone matrix. This may also explain the exceptional sizes of the site's porphyritic pitchstone blades, with those being the largest blades (the width commonly 20-30 mm) seen anywhere in Scotland in any material from any period. It is possible that, at Blackpark Plantation East, it was only feasible to produce blanks, which did not fracture around the frequently 5 mm large phenocrysts, by making the flakes and blades broader and thicker than contemporary counterparts in other, more fine-grained raw materials.

The Late Neolithic pitchstone assemblages include the same general tool categories as the Early Neolithic assemblages, that is, arrowheads, scrapers, piercers, plain knife forms (ie, backed and truncated pieces, as well as expedient scale-flaked pieces) and, not least, simple edge-retouched pieces. Like the Early Neolithic pitchstone assemblages (above), the Late Neolithic pitchstone assemblages include no serrated pieces. However, at Blackpark Plantation East, a well-executed, double-edged, finely serrated blade in flint was recovered (Ballin *et al.* forthcoming).

7. THE SOCIAL CONTEXT OF PITCHSTONE USE – NEOLITHIC TERRITORIES AND EXCHANGE NETWORKS

7.1 Introduction

As indicated in Chapter 1, one of the main purposes of the present report is to interpret the distribution of archaeological pitchstone across northern Britain, and not least its social context (Chapter 7.5). However, before this task can be carried out in a meaningful manner, it is necessary to provide a methodological, as well as terminological/conceptual background to the interpretation:

- As the distribution of archaeological pitchstone is somewhat biased by the ways the finds were recovered, the relationship between recovery, distribution and interpretation is to be briefly discussed in Chapter 7.2;
- A small number of relatively sophisticated approaches are essential to the manipulation of the data involved in the definition of the territorial structure and the pitchstone distribution patterns (Thiessen polygons, regression analysis and style/function analysis); these approaches are presented and discussed in Chapter 7.3;
- And, probably most importantly, it is necessary to briefly discuss the Neolithic setting of the pitchstone case (Chapter 7.4), as any reliable interpretation of the distribution of worked Arran pitchstone (within and beyond present-day Scotland), as well as the territorial structure and exchange network responsible for the dispersal of this material, must be based on an at least basic understanding of what a tribal society is.

In a sense, the exchange system can be perceived as the glue that allows the individual prehistoric territories to form a territorial *structure* by reinforcing kinship relations. Therefore, when a territorial structure for northern Britain has been proposed in Chapter 7.5.2, its 'binding' exchange system will be discussed (Chapter 7.5.2). An approach for the definition of prehistoric (tribal) exchange systems was developed, based mainly on Plog's 1977 paper *Modeling Economic Exchange* in Earle & Ericson's (1977a) *Exchange Systems in Prehistory*, but with later additions and amendments. In her master's thesis (2001) on *The Distribution of Geologic and Artefact Obsidian from the Silver Lake/Sycan Marsh Geochemical Source Group, South-Central Oregon*, Thatcher expanded Plog's methodology by including elements from mainly Skinner (1983, 87; 1997, 15), and the approach applied in the present analysis (Table 16) represents an adaptation of her approach.

The distinction between local, regional and exotic resources is based on Fisher & Eriksen (2002, 31, 71; the distances local/regional/exotic were originally defined by Geneste (1988a; b) as 'up to 5 km'; '5-20 km' and '30-80 km'), and the distinction between indirect, direct and embedded procurement on Morrow & Jefferies work (1989), which was inspired partly by Binford (1976; 1978).

Local raw materials or goods are defined as materials which

1	*Analytical focal point*: The type(s) of raw material or goods around which the analysis is centred.
2	*Duration*: The period of time during which this particular exchange system functioned.
3	*Magnitude*: The quantity of raw materials or items being exchanged.
4	*Diversity*: The variation of raw materials and artefact types present in the exchange system.
5	*Boundaries*: The geographical limits of the exchange system, and whether the exchanged raw materials/goods represent local, regional, or exotic resources.
6	*Acquisition*: Do the exchanged raw materials or goods represent indirect, direct, or embedded procurement?
7	*Directionality*: The flow of raw materials or goods in one or more directions.
8	*System shape*: Refers to the overall appearance and shape of the exchange system.
	8.1 <u>Symmetry</u>: The relative quantities of raw materials and goods flowing in one or the other direction.
	8.2 <u>Clustering</u>: Are the exchanged raw materials or goods abundant in particular areas?
9	*Object value*: Was the system concerned with the exchange of utilitarian or non-utiliarian raw materials or goods, or both?
10	*System complexity*: The amount of variation discovered within the exchange system. Uniform patterns of exchange indicate simplicity in a system, whereas a high degree of variation is characteristic of a more complex system.

Table 16. Characterisation of exchange networks (Plog 1977, with later additions).

were obtained less than 10 km from a given site, that is, within a traditional site catchment area (Higgs & Vita-Finzi 1972, 28); *regional* materials were obtained within 10 to 50 km from the site, that is, within the likely annual territory (Vita-Finzi & Higgs 1970; Higgs & Vita-Finzi 1972; Jarman 1972); and *exotic* materials were obtained more than 50 km from the site, that is, outside the annual territory. These concepts and definitions were clearly developed in connection with the analysis of hunter-gatherer societies, and they may need to be adapted to fit the pitchstone case.

Indirect procurement generally corresponds to exchange or trade; *direct* procurement to the acquisition of raw materials or goods by special purpose trips to the sources; and *embedded* procurement is defined as the acquisition of raw materials or goods within seasonal movements through the various economic territories (the catchment area and the annual territory). Like above, these concepts and definitions were developed in connection with the analysis of hunter-gatherer societies.

7.2 General Source Criticism – Recovery, Distribution, Interpretation

In this chapter, the relationship between the recovery, distribution and interpretation of Scottish archaeological pitchstone is dealt with in a summary fashion, and some general problems are highlighted. Some of these are examined further in Chapter 7.5, in connection with the discussion and interpretation of the territorial structure, and the pitchstone exchange network of Neolithic northern Britain.

When the distribution of worked Arran pitchstone (Figure 24) is scrutinized, some trends are obvious, such as the way sites yielding pitchstone seem to cluster. These trends probably represent prehistoric reality (Chapter 7.5), but there is little doubt that some of these trends have been exaggerated or subdued by a series of biasing factors. The most important of those are listed in Table 17.

7.2.1 The Industrial Era

The effect of the modern industrial era on the distribution of Scottish archaeological pitchstone corresponds to the general effect this era had on the distribution of prehistoric monuments. This phenomenon has been recognized throughout the industrialized world, where for example chambered cairns are practically non-existent in and around major cities (see distribution of dolmens and passage graves in Denmark, with the area of Greater Copenhagen representing an almost monument-free zone in a sea of Neolithic monuments; Jensen 2001, 365, 368). In Scotland, this effect is experienced in the Central Belt, with the zone between, and including, Greater Glasgow and Greater Edinburgh being considerably poorer in terms of Neolithic monuments than one would have expected (cf. various distribution maps in Henshall 1972).

Given the density of pitchstone-bearing archaeological sites around the inner Clyde basin (Figures 24-25), the relatively rich finds from recent excavations in the Stirling area (Chapelfield; Atkinson 2002) and in Fife (Balfarg Riding School; Barclay & Russell-White 1993), and the high land capability in the Central Belt (Davidson & Carter 1997, Fig. 4.2), which should have ensured a relatively high population density in this area in the Neolithic period, more archaeological pitchstone should have been recovered in the region between the Clyde and the Forth.

7.2.2 Population Density/Distribution in Modern Scotland

The relationship between the density and distribution of the present Scottish population and the recovery of archaeological pitchstone should be obvious: where the population density is extremely low, such as for example in the Scottish north, archaeological pitchstone is less likely to be found. The number of pitchstone artefacts on pitchstone-bearing sites (Table 3) suggests that there *is* a clear difference between the north and the south, with the former being relatively pitchstone-rich (Zone IIW and IISW have average pitchstone ratios per pitchstone-bearing site of 33% and 14%, respectively) and the latter relatively pitchstone-poor (Zone IV has an average pitchstone ratio per pitchstone-bearing site of 2%). It is, however, highly likely that the difference between the *total numbers* of pitchstone-bearing sites and pitchstone artefacts in the north and south is somewhat skewed, and that a higher population density throughout modern northern Scotland would have led to the recovery of more pitchstone objects.

Generally, it should be expected that, in prehistory, people would have been more evenly distributed throughout the

The industrial era
Population density/distribution in modern Scotland
Modern local development, and associated archaeological activity
The 'Tam Ward effect'
Archaeological preferences
Lack of awareness in pitchstone-poor areas
Erroneous lithic characterisation

Table 17. Factors affecting the overall distribution of prehistoric pitchstone artefacts.

country, with presently uninhabited, or almost uninhabited, zones having been the focus of some settlement or activity. The north-west is presently practically devoid of people, but would in prehistoric times almost certainly have included some fishing communities, or communities based on mixed farming and fishing. Probably due to the present low population density in the north-west and a much higher population density in the north-east, there is only one pitchstone find from the former (the enigmatic pitchstone pebble from Achnahaird Sands; Ballin 2002b), whereas several finds of actual pitchstone artefacts have been made in the area from Moray to Caithness. Similarly, the inner parts of northern Scotland would probably have carried a noticeable population along rivers and in valleys (as suggested by the pitchstone find from Urquhart Castle; Ballin 2005b), with associated shieling activities in more elevated areas. Recently, archaeological investigations of historical shielings in the Ben Lawers area (Atkinson *et al.* 2005) led to the recovery of several pieces of archaeological pitchstone, and it is highly likely that similar investigations in the north would lead to similar, albeit slightly less numerous, finds.

7.2.3 Modern Local Development

Although modern local development to a degree reflects the present population density of a region, there are variations in the level of local development which leaves pockets with curiously low densities of recovered archaeological pitchstone. One such area is Ayrshire, immediately south of Glasgow, where the counties of East, North and South Ayrshire used to form a low density area within a zone characterized by very high find frequencies (Figures 24-25). If the present project, and its discussion of pitchstone distribution, had been based entirely on finds available for examination through the main Scottish museums, this would have left a distinctly skewed picture, and it would have been very difficult, if not impossible, to explain this apparent 'recovery vacuum' on the eastern shores of the Clyde basin.

However, this distribution pattern was 'false', in the sense that it did not correctly reflect the prehistoric reality, and it was mainly a function of a lower than average level of modern development in this area. In recent years, many development projects have been carried out in the Ayrshire counties, and with the inclusion of finds from new excavations – partly newly excavated assemblages examined directly by the author at GUARD, University of Glasgow, and partly new finds picked up via *Discovery and Excavations in Scotland* – this 'vacuum' has more or less been filled. In distribution terms, the Ayrshire counties now form a logical part of the general distribution pattern, and this has allowed the construction of the basic zonation (Figure 4) suggested in connection with this report, as well as the proposed territorial structure and exchange network (Chapter 7.5).

7.2.4 The 'Tam Ward effect'

In connection with the compilation of the database of Scottish archaeological pitchstone, it became clear that various clusters and 'near vacua' might represent a human factor, such as, the geographical foci of individual amateur archaeologists and their different methodological approaches. Amongst some professional archaeologists, this has been coined the 'Tam Ward effect'. Tam Ward is the highly respected curator of the Biggar Museum Trust in South Lanarkshire and the energetic leader of Biggar Museum's Archaeological Group (BMAG). The BMAG was established in 1981, and it has organized repeated, systematic, and large-scale fieldwalking and excavation programmes on an annual basis, resulting in the recovery of almost 700 pieces of worked pitchstone (to 2008).

Over the years, the members of the BMAG have developed a degree of expertise, allowing them to securely recognize pitchstone. However, the group acknowledges the lay character of this knowledge, and has therefore established working relationships with relevant professionals, such as geologists and archaeologists. Within the field of archaeology, contacts have been made to specialists, who willingly advice the group in areas such as lithic/stone artefact characterization and analysis, petrology, pottery, and radiocarbon dating. As a consequence of the apparent concentration of pitchstone finds in the Biggar area, pitchstone recovery and research has developed into one of the group's core activities.

The question posed by archaeologists is whether this effect has seriously skewed the distribution pattern of archaeological pitchstone in southern Scotland, and whether for example the Scottish Borders would prove to be as find-rich as South Lanarkshire if scrutinized equally thoroughly? However, probably half of all pitchstone from the Biggar area was recovered during fieldwalking (the large assemblage from Corse Law entirely so; Clarke 1989), and although the fields of Tweeddale in the Scottish Borders have also been fieldwalked stringently (although not quite as systematically as in the Biggar area) they have not provided the same massive pitchstone assemblages (cf., Mulholland 1970). In the Scottish Borders area, the largest known assemblages include approximately a dozen pieces, and in most cases these collections were recovered during fieldwalking of many fields within a larger area (eg, labelled 'The Kelso Area' or 'Roxburghshire').

The huge concentration of archaeological pitchstone in the Luce Bay area may partly be an effect of local enthusiasts knowing that the sand dunes, and not least their deflation zones, might be rich in archaeological finds, and it is possible that this local knowledge has led to a slight over-representation of archaeological pitchstone in this area. However, the accumulation of pitchstone artefacts in the Scottish south-west is so massive, that there can be little doubt that the Luce Bay area truly was a focal point of the widespread exchange in pitchstone. It is interesting that a similar dune area in the east of Scotland (Culbin Sands in Aberdeenshire), where local archaeological enthusiasts have been combing the dunes and deflation zones equally thoroughly, has yielded practically no finds of archaeological pitchstone. Only four pieces have been registered as coming from this area, and according to Senior Curator Alan Saville at National Museums Scotland (pers. comm.), it is quite possible that these early finds have been mis-labelled and actually derive from the Glenluce Sands.

7.2.5 Archaeological Preferences

In the same way as some concentrations of pitchstone finds are due to the particular interests and activities of enthusiastic amateurs, some clusters of finds, as well as find vacua, may be explained by the specific activities and preferences (research agendas) of professional archaeologists. In southern and eastern Scotland, a number of important finds of pitchstone assemblages were made in connection with excavations carried out by Gordon Barclay from Historic Scotland, as part of his research into, for example, the settlements and ritual centres of the Neolithic period (eg, Balfarg Riding School, Claish, Nethermuir, North Mains, and Kinbeachie ; Barclay 1983; Barclay & Russell-White 1993; Barclay & Wickham-Jones 2002; Barclay *et al.* 2001; 2002).

In the north, there is a huge discrepancy between 1) the archaeological community's interest in Caithness compared to its interest in Orkney, and 2) the frequency of fieldwalking campaigns and excavations carried out in those areas, and the numbers of finds recovered north and south of the Pentland Firth. Orkney is understandably attractive to archaeologists, for a number of reasons. One is the island group's spectacular Neolithic settlements and burial/ritual monuments (eg, Foster 2006), but Orkney is also generally perceived as an area where the 'digger', can have a good time: due to the influx of substantially more tourists than in Caithness, south of the Pentland Firth, many more cultural activities take place, and there are many more options in terms of accommodation, restaurants, cafes, and pubs.

Although highly visible archaeological attractions, such as Skara Brae, Barnhouse, Maeshowe, the Stones of Stenness, and the Stones of Brodgar, suggest to the visitor that Orkney was a 'special place', the spectacularly uneven archaeological input experienced in the two neighbouring areas leaves the question open as to whether the great numerical difference between finds from Orkney and Caithness reflects prehistoric reality or, at least partly, a form of sampling bias?

7.2.6 Lack of Awareness in Pitchstone Poor Areas/ Erroneous Lithic Characterization

This factor generally relates to the problem of 'English pitchstone' (Ballin 2008a). The Anglo-Scottish border clearly represents a divide between a region where archaeological pitchstone is fairly common (southern Scotland) and where it is practically absent (England). Figures 24-25 show a number of distinct pitchstone concentrations near the Anglo-Scottish border, and a small number of finds south of the border indicates that the exchange network responsible for the dispersal of Arran pitchstone included at least parts of England. If one considers the fact that in Scotland archaeological pitchstone has been found 400 km north of the outcrops on Arran, it should be possible to recover the occasional piece of pitchstone at least as far south as Manchester. If one also considers the fact that the northwards dispersal of pitchstone probably largely stops where it does, due to the barrier created by the Atlantic Ocean, worked pitchstone could hypothetically have travelled as far south as the English Channel.

Until recently, only three finds were known immediately south of the border: one from Bowden Doors in Northumberland (Burgess 1972) and two from Carlisle in Cumbria (Fell 1990). In connection with the present project, contacts were made with colleagues in Northumberland and Cumbria, and focused fieldwalking in the former county has now led to the recovery of 15 pieces of worked pitchstone from a limited number of neighbouring fields at Low Trewhett, near Rothbury (Kristian L.R. Pedersen pers. comm.). It is highly likely that the almost total absence of archaeological pitchstone in England is the result of this material not being expected south of the border, whereas, today, every Scottish archaeologist knows of Arran pitchstone and the possibility that pitchstone artefacts may be found throughout the country (not least after the recovery of 23 pieces at Barnhouse on Orkney; Middleton 2005; Richards 2005).

This difference in perception has led to the fact that Scottish archaeologists are trained in the recognition of pitchstone, whereas English archaeologists are not. Subsequently, archaeological pitchstone found south of the border (where it is probable to appear infrequently, and in small numbers, like in most of northern Scotland) is thought to have been characterized erroneously as materials more commonly found in English archaeological contexts, such as dark forms of chert and flint, as well as jet and glassy slag (Chapter 3.6). This problem was highlighted in an issue of PAST, the popular magazine of the Prehistoric Society (Ballin 2008a), and pitchstone samples have been distributed amongst the members of the Implement Petrology Group, in the hope that this may lead to pieces of 'English pitchstone' being noticed and reported.

7.2.7 Conclusion

Although, as described above, the present distribution of Scottish archaeological pitchstone may be slightly biased, due to a number of distorting factors (Table 17), it is nevertheless the author's impression that the distribution map Figure 24 shows genuine trends of relevance to the interpretation of the territorial structure and the pitchstone exchange network of Neolithic northern Britain. Any flaws relate to detail rather than general principles. Presently, the most important weakness is the uncertain situation south of the Anglo-Scottish border, which makes it impossible to estimate the true size and complexity of the pitchstone exchange network. This is to be discussed further in Chapter 7.5.

7.3 Distribution Analysis – Some Key Approaches

7.3.1 Thiessen Polygons

Orton (1980, 192) defines Thiessen polygons as polygons which, on a map, '... are constructed simply by drawing lines at right angles through the mid-points of the lines joining neighbouring centres. Every point in [an] area is nearer to its centre than to any other [centre]'. They have been used to define the administrative structure of Romano-British walled towns in central and southern England (ibid., Fig. 7.7), as well as the territorial structure of Early Neolithic Arran (Renfrew 1976, Fig. 6).

There is no doubt that Thiessen polygons are relatively uncomplicated to use when applied to a highly structured society like Roman Britain, where administrative units were defined and run according to central decrees, but the situation is slightly more complex in prehistoric societies. In prehistoric societies, territories were defined by a multitude of constantly changing factors, some relating to natural factors, and some to human factors. One of the main problems, when applying Thiessen polygons to prehistoric societies, is that ideology may differ markedly between neighbouring territories, and that, subsequently, one or the other group may express territoriality in a way that leaves that territory/those territories 'invisible' to the analyst. The consequence in such cases is that the territories bordering on to these 'invisible' territories may appear larger than they actually were in the prehistoric reality.

If, for example, it was attempted to analyse the territorial structure of south-west Scotland simply by drawing Thiessen polygons around Clyde and Bargrennan type tombs, this would produce skewed results as, apparently, some groups preferred non-megalithic long cairns (eg, Cummings 2002, Fig. 9; also Telford 2002). The territories of the latter groups would simply 'disappear'. The same phenomenon can be experienced in other parts of Scotland, where an area may be dominated by a particular type of megalithic or non-megalithic burial monument, but where local groups, on occasion, chose to express their territoriality in deviating ways (cf. various distribution maps in Henshall 1963; 1972).

Although there can be little doubt that local prehistoric groups on occasion deviated from the regional norm, some variation may also be chronological. If the burial

monuments of prehistoric Scotland are used as an example yet again, it is obviously important when applying Thiessen polygons, that these are drawn around contemporary monuments, which are territorial markers within the same territorial structure. As prehistoric territorial borders may have fluctuated somewhat (Parkinson 2002, 9), one cairn could theoretically have replaced another, for example in connection with border corrections. To remedy this problem, Renfrew (1976, 211) suggests to use a combination of solitary cairns and clusters of cairns as markers.

Another important point in connection with the use of Thiessen polygons is the question of message recipients, as touched upon in Chapter 7.3.3. As territorial markers are expressions of style (that is, '...formal variation in material culture that has a distinct referent and transmits a clear message to a defined target population about conscious affiliation or identity'; Wiessner 1983, 257), they would not have been used to transmit a message to the closest relatives or friends, as these already knew the message (Figure 21).

In his paper on Neolithic territorial structures, Renfrew (1976) sub-divides the islands of Rousay in Orkney (ibid., Fig. 4) and Arran (ibid., Fig. 6) by the use of Thiessen polygons. As the resultant 'territories' are exceedingly small (on average from 1-6 km across), those are more likely to be Neolithic farmsteads than territories *sensu stricto* (the latter would represent larger conglomerates of people such as, in the Neolithic period, clans or tribes; Chapters 7.4.1-2). Although each cairn may mark the land-rights of a particular family, or lineage, the more important message of the cairns may be directed outwards, outlining Rousay and Arran, in their totalities, as the territories of specific clans or sub-clans. This proposition may be supported by the strictly coastal location of the Rousay cairns.

Obviously, the territories suggested by the Thiessen polygons do not correspond exactly to the prehistoric territories – they are approximations, or mathematical abstractions. However, they may be helpful in suggesting how *many* territories might have existed within a particular region at a specific time, as well as in suggesting the approximate sizes of these territories. As clearly indicated by Norwegian research (eg, Alsaker 1987, 59), topographical barriers – such as fiords, rivers and mountain ranges – frequently served as territorial boundaries (eg, O'Shea & Milner 2002, 208), and it should be possible to correct the territorial structure suggested by the Thiessen polygons by comparison with the region's topography.

7.3.2 Regression Analysis

Regression analysis is the statistical technique by which the relationship between two variables is studied (Orton 1980, 116). In archaeology, it has traditionally been applied as part of the analysis of exchange systems, and it became more widely used from the mid-1960s, when Renfrew *et al.* (1966; 1968) combined it with raw material provenancing to describe the procurement patterns of the Near Eastern obsidian exchange. In this chapter, regression analysis is discussed to the degree it was found relevant to the present debate on prehistoric pitchstone in northern Britain (for a more detailed overview of the method's mathematical and statistical background, consult Renfrew 1977; Hodder 1974; Orton 1980, 116). Exchange is here defined as in Renfrew (1977, 72), that is '... in the case of some distributions it is not established that the goods changed hands at all; [exchange] in this case implies procurement of materials from a distance, by whatever mechanism'.

In the archaeological literature which deals with regression analysis *per se* (ibid.), a multitude of examples are given (many regarding the distribution of artefacts or goods in more complex later prehistoric or early historic societies), and a large number of sophisticated interpretational models are presented. However, it should be borne in mind, that the northern British exchange in lithic and stone artefacts generally took place in the earlier part of prehistory and in tribal societies (Chapter 7.4.1). The present presentation of regression analysis has therefore been restricted to examples perceived as relevant to pitchstone exchange, and to interpretational models potentially relevant to this topic.

When analysing early prehistoric exchange, regression analysis usually results in a diagrammatic presentation, that is, a diagram with two axes, each representing one of two variables, and a regression line (a fall-off curve), showing how one variable changes as the other is varied (Orton 1980, 116). Traditionally, the x-axis represents distance to source, and the y-axis quantity. Quantity usually decreases with distance to source, and the specific way in which it decreases may provide information on the prehistoric society in question, as well as on the particular character of the exchange network responsible for the artefact distribution.

Generally, only a small number of regression forms (or forms of fall-off curves) are encountered in connection with the analysis of early prehistoric exchange of lithic raw materials or artefacts. They are:

- A gradually sloping fall-off curve, which may take the form of either a straight line or an exponentially declining curve; traditionally, this form has been interpreted as representing so-called down-the-line exchange based on reciprocity (eg, Renfrew 1977, 77);
- A multi-modal fall-off curve (ie, a curve with one or more marked peaks); in connection with the interpretation of early prehistoric societies, this form is usually perceived as representing directional exchange, involving central places, redistribution, and a degree of social control (eg, Renfrew 1977, 85) – in connection with later, more complex societies this form may indicate central place market trade (Torrence 1986, 116);
- A stepped fall-off curve, where the steps represent the

borders of various social territories; in this model, the exchanged artefacts or raw materials were imbued with stylistic meaning, and attempts were made to distribute the artefacts/raw materials evenly throughout the territory, as possessing a specific proportion of the artefacts/raw materials in question defined the owner(s) as belonging to a specific social group.

Regression analysis would obviously be complicated by the existence of competing raw material resources, which may introduce a level of interference into any diagrammatic presentation (Ericson discusses a case where a group simultaneously exploited three different obsidian quarries; Ericson 1982, 144), and the following examples are all based on raw materials with strictly localized sources.

The distribution of Cambrian flint (or Kinnekullen flint) in central Sweden is an eminent example of the first form of regression. This type of flint, which was used and exchanged in the Late Mesolithic Lihult period of western and central Sweden, is characterized by conchoidal fracture as well as numerous cracks, and it is generally perceived as a relatively poor raw material (Kindgren 1991; Högberg & Olausson 2007, 132). It is found at the Kinnekullen peak, near the

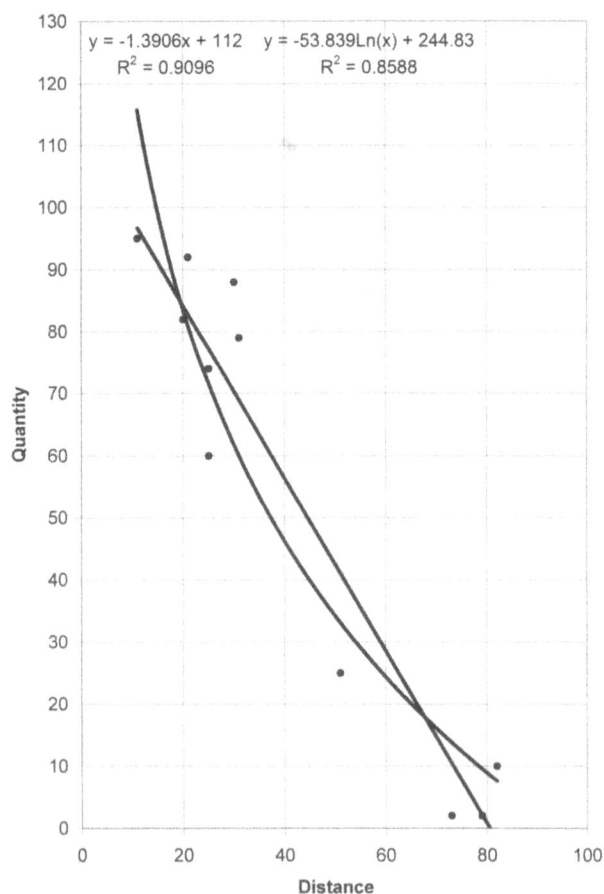

Figure 18. Linear and logarithmic fall-off curves for Cambrian flint from the Kinnekullen peak in central Sweden (based on information from Kindgren 1991, Table 4). The correlation coefficients of the two curves have been calculated.

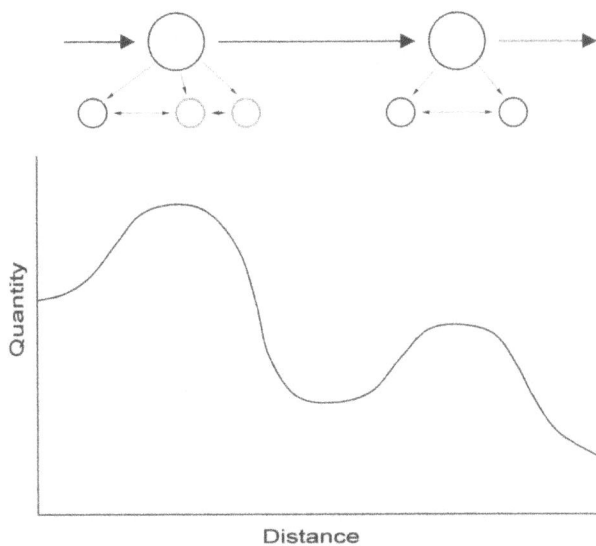

Figure 19. Simplified model of directional exchange (after Renfrew 1977, Fig. 5).

obtainable at a lower price (if obtained by exchange) or input of labour (if obtained by direct procurement). The exaggerated steepness of the curve is an indicator of the relatively low value of this resource.

Due to the mostly limited complexity characterising early prehistoric (tribal) societies, directional exchange (Figure 19) is rarely experienced in connection with the distribution of lithic raw materials and artefacts. However, it is thought that, frequently, the exchange in axeheads (and possibly other more elaborate artefacts and prestige goods) was directional, and that they may have been exchanged in bulk. Cornish Group I axeheads and Cumbrian Group VI axeheads have been mentioned in this context, as both types of axeheads share a distribution pattern characterized by dense concentration in the area around the stone quarries, followed by a relatively thinly covered zone, and further away from the outcrops remote clusters of axeheads are found (Bradley & Edmonds 1993, 45, Fig. 3.1).

It has been suggested, that the clustering of Cornish axeheads in the Thames estuary may indicate seaborne exchange (Cummins 1979, 10), whereas the Cumbrian axeheads would have had to be transported over land (possibly including river transport) to their secondary centre in Lincolnshire. Large hoards of Danish Middle Neolithic flint axeheads found in northern Sweden (Becker 1952) are other examples of directional exchange and, like in the case of the Cornish axeheads, these implements must have been transported by sea and in bulk. The northern British pitchstone exchange is thought to represent another example of directional exchange (see fall-off curve in Figure 27).

Hornborga Lake, from where it was distributed throughout a zone with a 100 km radius (Figure 18). The fall-off curve for this resource, whether linear or logarithmic, is gradually (although steeply) declining, and both curves are associated with a high correlation coefficient, showing an almost perfect relationship between distance and quantity (R^2 = 0.91 and 0.86, respectively). It is clear that, in this case, a particular low-grade raw material was exploited heavily near the source but, with growing distance to source, it was gradually replaced by other raw materials which were

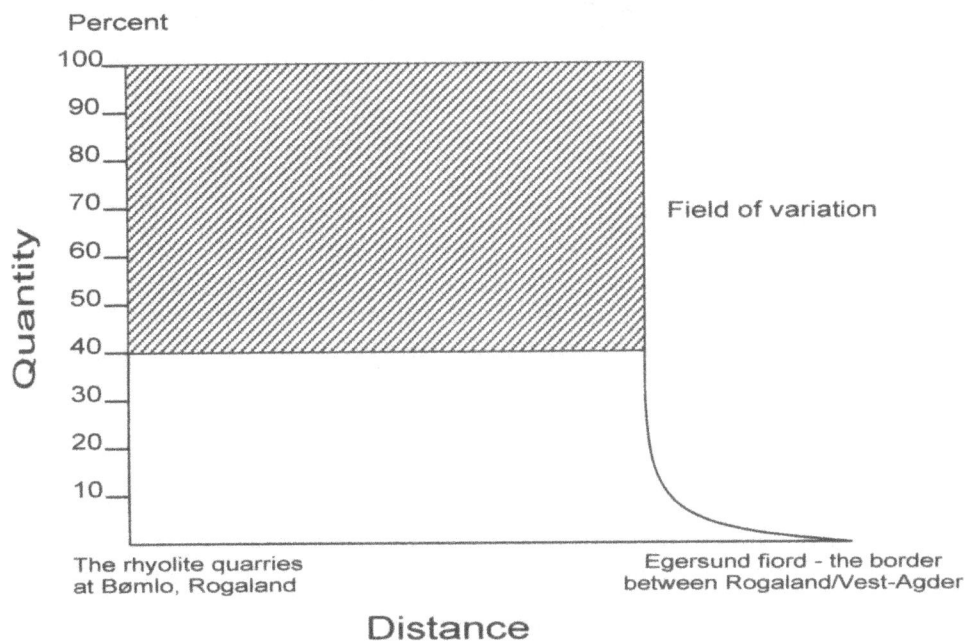

Figure 20. Idealized fall-off curve for the distribution of rhyolite from the quarries at Bømlo, south-west Norway. Within this specific social territory the rhyolite proportion of Early Neolithic assemblages varies between 100% and 40%, probably due to different assemblage dates, with earlier assemblages being characterized by close to 100% rhyolite, and later assemblages somewhat less. At the border of the south-west Norwegian social territory, the rhyolite proportion drops drastically to near 0%.

A stepped fall-off curve may be based on direct procurement as well as directional exchange and redistribution. However, where the exchanged raw material described by a gradually declining fall-off curve represented function, and a multi-modal fall-off curve either function or style, exchanged raw materials characterized by a stepped fall-off curve were almost certainly perceived in a stylistic light, that is, they defined any owner as belonging to a specific social group.

The south-west Norwegian exchange in rhyolite is an example of a stepped fall-off curve (Figure 20). It is very difficult to produce an actual regression line describing this exchange, as many of the region's settlements are chronologically mixed, but it is clear that 1) the procurement of rhyolite from the quarries at Bømlo in central coastal Rogaland started around 5200 BP (more or less exactly at the transition between the Norwegian Late Mesolithic and Early Neolithic periods) (Alsaker 1987, 97), 2) from this time and through the Early Neolithic period, south-west Norwegian assemblages are dominated by rhyolite use, supplemented by some use of flint and quartz (Bruen Olsen defines 'dominance' as 40-85% of the lithic material; 1992, 85), and 3) the frequency of rhyolite drops exceedingly abruptly at the borders of the perceived south-west Norwegian social territory (Alsaker 1987, 97), as this territory has been defined by the distribution of greenstone axeheads (Bergsvik & Bruen Olsen 2003, Fig. 52.6).

In topographical terms, the south-west Norwegian social territory is defined in the north by the deep Sognefjorden; in the south by the Egersund fiord; and in the east by the Norwegian High Mountains (Alsaker 1987, 59). In 1992, Oslo University carried out excavations on the Lista peninsula, immediately outside the southern border of the south-west Norwegian social territory, and although *c.* 700,000 lithic artefacts were recovered (Ballin & Lass Jensen 1995), including material from all phases of the Mesolithic and Neolithic periods, only two refitting fragments of the same rhyolite arrowhead were found.

In north-eastern lower Michigan, the distribution of Bayport chert in the Juntunen Phase is also characterized as stepped, and the authors note that the various steps correspond to the marked territories of Juntunen bands (O'Shea & Milner 2002, Fig. 6b). However, in addition to the territorial factor, the distribution of Bayport chert is also influenced by the sites' relative distance to the shores of Lake Huron which must, in O'Shea & Milner's words (ibid., 220), represent an 'ease of transport' effect produced by canoe travel on the lake.

7.3.3 Style and Function

The concept of *style* was introduced in ethnographic and ethno-archaeological circles around 1970, and it was adapted to archaeological contexts firstly by Martin Wobst (1977) and Polly Wiessner (1983; 1984) (for a research-historical discussion, see Gendel 1984 and Gebauer

1987). The study of stylistic variation has always been an important, albeit implicit, ingredient in the construction of chronologies but with New Archaeology's focus on explanation of observed variation the concept of style became dynamic.

The conception of stylistic variation as observed in the present volume is based on Wobst's definition of style as exchange of information (Wobst 1977, 317). This perception is elaborated on by Wiessner, who defines style as '...formal variation in material culture that transmits information about personal and social identity' (Wiessner 1983, 256). Wiessner distinguishes between two forms of style, with one form relating to personal identity (assertive style), whereas the other form relates to social or group identity (emblemic style) (Wiessner 1983, 257). Assertive style is of no relevance to the present paper and it is not discussed below. Wiessner defines emblemic style as '...formal variation in material culture that has a distinct referent and transmits a clear message to a defined target population (cf. Wobst 1977, 323) about conscious affiliation or identity' (Wiessner 1983, 257); emblemic style (in the following text abbreviated to 'style') functions as identification (within-group) as well as differentiation (between-group) (Wobst 1977, 327; Hodder 1979, 447; Wiessner 1983, 257). Wobst (1977, 321) and Wiessner (1983, 261) also define style negatively, that is, as variation not due to function, raw-materials and technology.

Analyses of stylistic variation relating to territoriality and social organization have shown that style is primarily displayed on one specific spatial/social level, namely the level of the social territory (Helm 1973: *the tribe*; Wiessner 1983: *the language group*). The main reason for this is the fact that style, as a medium for social information, has an economic side, and it is only displayed when information cannot be exchanged in simpler fashions, for example verbally (Wobst 1977, 323) (Figure 21). Thus, stylistic behaviour will increase gradually with the increasing size and complexity of the social network. In smaller social networks, like bands, there is not the same need to express social identity via style, and in a techno-complex the members of that territorial entity feel little mutual identity and therefore express themselves stylistically to a lesser extent. The techno-complex is '...a larger and looser entity than the culture group; an entity of larger size but lower rank than either the culture or culture group' (Clarke 1968, 321). More precisely, the techno-complex is a territory based primarily on a common economic approach and to a lesser degree social identity, with its main material manifestation being function rather than style.

Raw-materials and technology have, like function, been perceived as opposites to style (Wiessner 1983, 261), but when inferring territoriality and social organization the situation is more equivocal; thus, it is the author's view that raw-materials and technology may operate as stylistic as well as functional expressions.

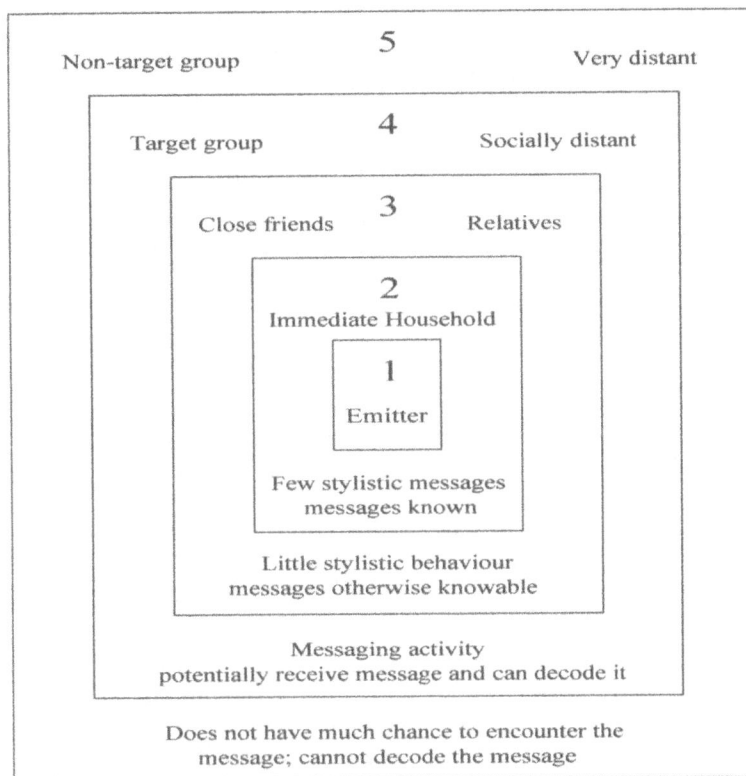

Figure 21. The stylistic message in relation to various target and non-target groups (Wobst 1977, 325).

If a decision to use or not use a certain raw-material is based entirely on the presence or absence of this raw-material the expression is functional, whereas a decision to give preference to a rare raw-material, or a decision to disregard a suitable abundant raw-material, are stylistic expressions (exchange of social information). The total dominance of quartz on, for example, Shetland is probably an example of the former, as practically no other suitable raw materials were available for the manufacture of the prehistoric Shetlanders' everyday tools, whereas the use of specific raw-materials in the axehead production of West Norway is an example of the latter, with raw-material preference not reflecting raw-material availability (Bruen Olsen and Alsaker 1984, 96). In the latter case, two extensive social territories were supplied almost completely from one central quarry within each of the two territories.

Raw-material preference as an expression of function usually results in a gradually declining *fall-off curve* (Renfrew 1977, 73) with growing distance to the outcrop, whereas raw-material preference as an expression of style results in a marked drop in frequency at the borders of the social territory in question (Hodder 1979, 447), or possibly a stepped decline (O'Shea & Milner 2002, 220). Fall-off curves are explained and discussed further in Chapter 7.3.2.

Technology expresses function in cases where the choice of technology was based entirely on practical considerations relating to the manufacture of specific products (cost-benefit), but it expresses style in the cases

where one technology or technique was chosen from a host of possibilities each capable of delivering the same product at the same price. The introduction of the quirky Levallois-like technique at the beginning of the British Late Neolithic period may be an example of the former: most probably the main function of this technique was to allow the production of broad flakes for chisel-shaped arrowheads (from the Levallois-like core's main flaking-front) and slender blades for cutting implements (from the core's flanks) from the same parent pieces, that is, the decision was based on practical considerations. The introduction in the Late Kongemosian / Early Ertebølle Culture of Jutland of the so-called *Kombewa technique* (Andersen 1978; Inizan *et al.* 1992, 57; Ashton *et al.* 1991) is an example of the latter, as the intended bi-convex flakes (ie, flakes with two opposed ventral faces) may have been desired mainly for their appearance, and these implements may have been carriers of stylistic information, allowing people in southern Scandinavia to distinguish between groups from the west (bi-convex arrowheads) and groups from the east (arrowheads based on traditional ventral-dorsal flakes).

7.4 Basic Theoretical Framework

7.4.1 Archaeology and Ethnography

As mentioned in Chapter 7.1, any attempt at explaining the distribution of Scottish archaeological pitchstone must be founded on an understanding of what a Neolithic or tribal society is, and how it was structured and worked. To achieve this, it is essential that the archaeologist works with the ethnographer, and Carneiro (2002, 49) suggests to unite the two within the field of anthropology, with archaeologists and ethnographers being '... full and trusting partners'.

There are different views on the strengths and weaknesses of the two disciplines, archaeology and ethnography, and most analysts probably perceive ethnography as the stronger of the two, as ethnographers deal with complete living societies, whereas archaeologists deal with truncated material cultures, on the basis of which they have to deduce a prehistoric reality (Carneiro 2002, 49). Fowles (2002, 14), on the other hand, emphasizes archaeology's longer time perspective, and he characterizes the ethnographic record as the more static of the two, as it is concerned with what he refers to as 'short-term events': through the totality of time, the spectrum of society formation and human behaviour probably included solutions which do not exist today, as they represent adaptations to concrete, no longer existing situations – 'Just as a day in mid-summer will not serve as a model for an entire year, neither can a purely ethnographic model stand for an archaeological one' (ibid.).

This view is shared by Earle & Ericson (1977b, 9) who emphasize the dangers of blindly applying ethnographic parallels, as most so-called primitive societies today are influenced by contact through colonialization and post-colonial development. However, no matter whether the individual analyst favours one or the other of the two sister disciplines, one should bear in mind Fowles' (2002, 23) words (quoting Braudel 1980, 50): 'The point at which models (...) come to be universally applied in law-like form is the point at which we cease to explain or provide insight into social phenomena'.

The following chapters (7.4.2-4) do not represent a *discussion* of the ethnographic concepts and terminology relating to tribal kinship-structures, territories and exchange networks, but merely a *summary* of the general consensus within this field. In Chapters 7.4.2-4, certain ethnographic cases are given more weight than others (eg, the Hopewell Interaction Sphere; Caldwell 1964; Struever & Houart 1964; 1972; Struever 1972; Seeman 1995; Yerkes 2002), as they are perceived to be more relevant to the specific issue of pitchstone distribution and exchange.

7.4.2 Tribal Society and its Kinship Structure

'Tribe', as the term is most commonly used today, was defined by Service (1971) in his ground-breaking work on *Primitive Social Organization*. In his book which is inspired by earlier work by, among others, Radcliffe-Brown (1948), Evans-Pritchard (1940), and Steward (1955), he proposed an evolutionary sequence of cultures, which he labelled bands, tribes, chiefdoms, and states (Table 18).

Service's tribal concept was adopted by Sahlins (1961, 93), who suggested the following definition of the term:

'A band is a simple association of families, but a tribe is an association of kin groups which are themselves composed of families. A tribe is a segmental organization. [...] It is sometimes possible to speak of several levels of segmentation [...]. A "primary tribal segment" is defined as the smallest multi-family group that collectively exploits an area of tribal resources and forms a residential entity all or most of the year [...]. Small localized – often primary – tribal segments tend to be economically and politically autonomous. A tribe as a whole is normally not a political organization but rather a social-cultural-ethnic identity. It is held together primarily by likenesses among its segments [...] and by pan-tribal institutions, such as a system of intermarrying clans, of age grades, or military or religious societies, which cross-cut the primary segments. Pan-tribal institutions make a tribe a more integrated social institution [...] than a group of inter-marrying bands [and they] are perhaps the most indicative characteristic of tribal society.'

Most commonly, tribes are based on cellular social sub-division, with the core kinship units being extended families, lineages, and clans (Carneiro 2002, 41-2). A *lineage* consists of a number of extended families, who are related either through the male or female line. A *clan* consists of a number of lineages, the members of which all consider themselves interrelated, but who may not be able to trace their genealogies back to an actual common ancestor; frequently, the founder of a clan is a more or less mythical figure, who might even have developed into an

	Band societies	*Tribal societies*	*Chiefdoms*	*States*
Social organization	Egalitarian	Segmentary society	Kinship-based ranking under herediatry leader	Class-based hierarchy under king or emperor
	Informal leadership	Pan-tribal associations		
		Big Men'		
Economic organization	Mobile hunter-gatherers	Settled farmers	Centralized accumulation and redistribution	Centralized bureaucracy
		Pastoralist herders		Tribute-based
			Some craft specialization	Laws and taxation

Table 18. The social and economic organization of band societies, tribal societies, chiefdoms, and states (Service 1962); adaptation of definitions given in Renfrew & Bann (1996, 167).

animal (a totemic ancestor), whom clan members were forbidden to kill or eat (taboo). A number of clans may then form a *tribe* (frequently a loosely joined linguistic group; eg, Morgan 1995, 93; Radcliffe-Brown 1948, 23) which, on occasion, might be united in large *tribal confederacies* (eg, Evans-Pritchard 1940, 5). All of the above social units may be either sub-divided (sub-lineages, sub-clans) or joined in clusters. Pan-tribal sodalities (mostly non kin-based organizations) include age grade associations and secret societies, but they may be concerned with all forms of activities, such as ritual and warfare (Service 1971, 102).

Although some social groups may easily be defined as lineages, clans, or tribes, there are cases where this is less straightforward. Carneiro (2002, 50) mentions a case of nine neighbouring villages in the Upper Xingu region of central Brazil. They represent three very different linguistic groups, but they also form part of a mutual exchange network, inter-marry, compete in sporting events, and hold joint ceremonies. Yet, the Amazonian ethnographers working in the region do not refer to this group of villages as forming a tribe.

Some analysts have discussed why 'tribalization', the process by which tribes came about (Anderson 2002, 248), happens at all. Braun & Plog (1982) suggest that 'tribalization' occurs as a strategy to minimize risk and overcome subsistence stress, shortfalls, and general uncertainty. Whether the problem at hand was natural (droughts, floods, pests, diseases, etc.) or, for example, related to aggressive neighbours, the formation of alliances was one solution, and *usually* these alliances would be kinship-based. Bender (1985) points out that 'tribalization' allowed alliance formation at a larger scale than that made possible by band interaction. However, in terms of security issues, alliance formation is key and trumps kinship: as stated by Dalton (1977, 194), Kwakiutl killed other Kwakiutl who were not their allies as readily as they killed non-Kwakiutl who were not their allies.

In addition to its segmented structure and pan-tribal institutions, a tribe is defined by a specific leadership structure, which clearly distinguishes it from band societies and chiefdoms (Table 18). A tribal leader is commonly referred to as a 'Big Man' (a term adopted from Polynesian tribal societies), and he is more influential than the *ad hoc* activity leaders of band societies, but his position does not depend on birth, as is the case in chiefdoms. In tribal societies, leadership positions are won by demonstrating superior ability, achievement, and luck, in fields such as oratory, settling of private disputes, war prowess, peace negotiation, the securing of strong allies, and the organization of ceremonial exchange (Dalton 1977, 196).

Srivastava (2008) defines a Big Man as a tribal leader whose role is like that of a village headman, but whose authority is regional, although the term is generally used for village leaders as well as regional leaders. Part of the

definition of a Big Man is that he must be generous, and that he must work hard to generate a surplus, which can be distributed amongst his followers and thus converted into prestige (ibid.). The consequence of this tribal form of redistribution is that (in contrast to chiefdoms) the leader, the Big Man, is frequently the least wealthy person in the village or clan: 'The headman (ie, Big Man) of a Lengua village in the Paraguayan Chaco, for example, is likely to be the poorest man in the village since, if he happens to acquire some material possessions which someone else lacks, he is expected to part with it' Carneiro 2002, 40).

Although the terms band, tribe, and chiefdom are helpful when discussing the evolution of socio-political structures in primitive societies, it is more or less agreed amongst ethnographers that there are many hybrid organizational forms, and that the evolution of social and political organization may have been more fluctuating than continuous or stage-based (Fowles 2002, 19; Figure 22).

7.4.3 Tribal Territories

Due to the segmented character of tribal society, where each kinship-level may hold land (eg, O'Shea & Milner 2002, 207), a tribal territorial structure is usually considerably more complex than that of a band society. Based on his work on the Mesolithic period of Scandinavia, Clark (1975,

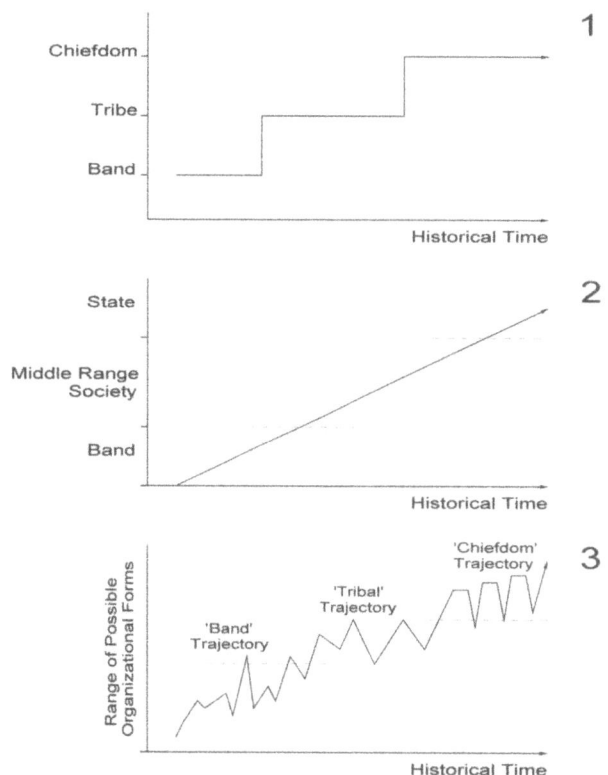

Figure 22. Three evolutionary models: 1) stage-based or ideal-typic model of general evolution; 2) continuous model of evolution (social types are arbitrarily defined as ranges of organizational variability); and 3) 'zzz-evolution' as types of social trajectories. According to Fowles (2002, Figs 1-3), the third model may most accurately describe the fluctuating evolution of human societies.

69) suggested a four-level socio-economical territorial hierarchy, which included 1) the catchment territory, 2) the annual territory, 3) the social territory, and 4) the techno-complex.

The catchment territory is the area exploited daily from the individual base-camp. In the case of hunter/gatherers with no mechanical means of transport this normally constitutes an area with a radius of approximately 10 km (Vita-Finzi & Higgs 1970; Higgs & Vita-Finzi 1972; Jarman 1972). Clark defines *the annual territory* as the area exploited by a mobile band over the course of a year. A band is rarely able to rely entirely on one eco-zone and has to be mobile to exploit the fluctuating riches of the seasons and the landscape. A detailed description of an annual territory is presented in Binford (1983, 109). *The social territory* is the area drawn upon by the band for raw-materials, finished products and mates; this takes the form of *systematic* exchange between groups with kinship ties. According to Clarke (1968, 291), this level constitutes '...an area of maximal internal intercommunication and diffusion', and, according to Wobst (1974, 152) '...the highest level of social integration'. Internally, the social territory is defined by a relatively uniform material culture expressing social unity, thereby ensuring a feeling of basic daily security. The *techno-complex* is composed of a number of social territories between which there is usually a low degree of social interaction. The social territories forming part of a techno-complex share essential functional artefact types and, first of all, technological elements based on a common general subsistence strategy, usually geographically delimited to an area with a uniform topography and ecology.

As most tribal groups are sedentary (although some nomadic groups are also tribal), Clark's first two levels, the catchment territory and the annual territory, are of little relevance to the discussion of tribal territorial structures. In general, tribes are defined by a hierarchy of social territories, each linked to a specific kinship level (O'Shea & Milner 2002, 207). These different levels of social territories may be characterized by markers in the landscape, as well as by material culture markers (style; Chapter 7.3.3). If a landscape was poor in terms of natural markers (rock outcrops, summits, river mouths, points, etc.), markers would be built. Built markers could be boundary cairns, or burial or ritual monuments (ibid., 2008; also Goldstein 1976). Material culture markers (ie, stylistic elements) were commonly arrowhead or axehead forms, ceramic styles, or 'prescribed' raw material preferences (cf., Anderson 2002, 267).

As mentioned in Chapter 7.3.2 (regression analysis), territorial boundaries were commonly characterized by an abrupt drop in the frequency of stylistic elements (ie, a stepped fall-off curve). As put by O'Shea & Milner (2002, 217) in their analysis of the raw material distribution in the Juntunen Phase of the upper Great Lakes region of the Unites States:

'... we expect the proportion of a given raw material present in an assemblage to be inversely related to the distance from its source, and to be positively related to its quality. Deviations from these expectations may (among other things) reflect the presence of social boundaries that may impede or facilitate distribution'.

A number of factors occasionally complicate the analysis of tribal territorial structures, such as the fact that a clansman may have had access not only to his own clan estate, but also to the estates of his wives and his mother (Blundell & Layton 1978, 239). It is another complicating factor that some groups exhibit clear boundaries, whereas other territories appear as '... smears across the [...] landscape' (Parkinson 2002, 8). While some boundaries seem to have been fixed through millennia (for example the boundary between the West-Norwegian stone axehead complexes; Bruen Olsen & Alsaker 1984; Bergsvik & Bruen Olsen 2003), many tribal boundaries were in constant flux (Parkinson 2002, 9).

7.4.4 Exchange and Exchange Systems

It has been common practice to interpret Stone Age economics within the framework of modern economic mechanisms, as demonstrated by archaeological terminology. It is, for example, still custom to refer to the exchange in stone axeheads as 'trade' (eg, Bradley & Edmonds 1993), and the axehead production sites are referred to as 'axe factories' (Houlder 1979). Both terms imply a level and form of organization, as well as driving motives (profit), which were all but unknown to Neolithic man, and they should be avoided.

In contrast to trade, exchange usually has a distinct social side, namely that of regulating interaction between groups of people ('Exchange may be looked upon as a form of social communication'; Wilmsen 1972, 1). The overarching principle behind exchange is simply to provide security, in material as well as social form. The material basis of exchange is the fact that different tribal groups, within different territories, had variable access to natural resources, and via participation in exchange networks these resources were distributed throughout the region, and possibly beyond. The form of the exchange forged alliances and helped to maintain the peace or, in the case of threats from outsiders, alliance partners could either be asked to help in the war effort, or they could provide a safe haven to which a group could flee, should they be overrun ('Regional trade is a form of foreign policy; Ford 1972, 43).

In most situations, exchange is based on kinship (eg, Renfrew 1993, 11), although the formation of non-kin 'gift-partnerships' is also common (Orme 1981, 180). The kinship-basis of tribal exchange, and the varying forms of this exchange, is most clearly explained by Sahlins (1972). He suggested that tribal exchange generally takes one of

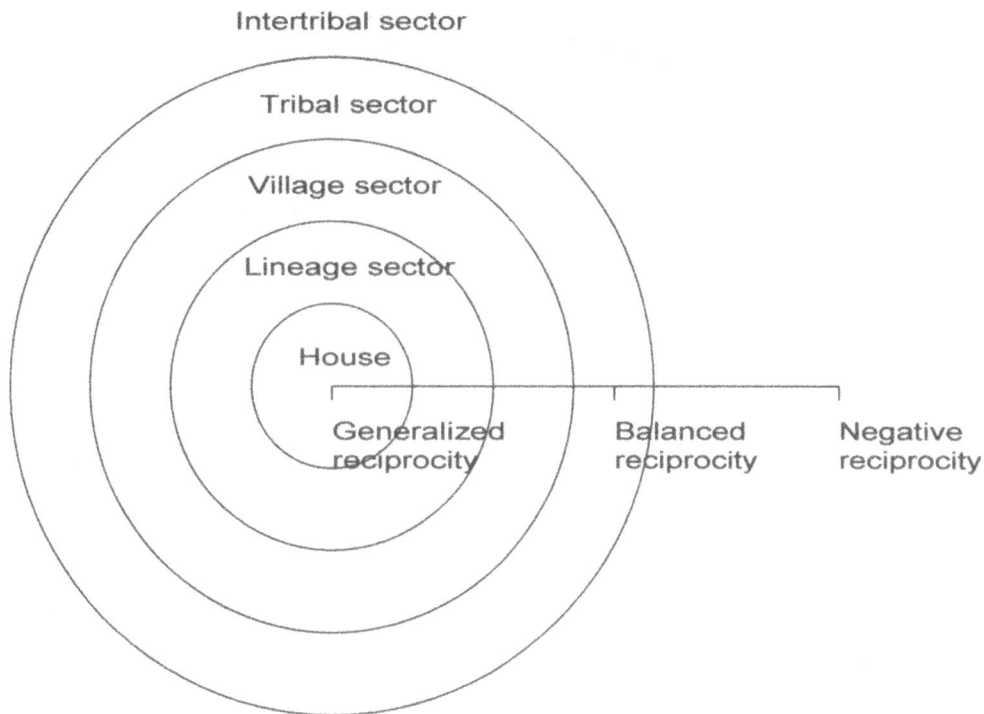

Figure 23. Reciprocity and kinship residential sectors (after Sahlins 1972, Fig. 5.1).

three forms: generalized, balanced or negative reciprocity. He defined *generalized reciprocity* as transactions that are putatively altruistic, that is, where assistance is given and, if possible and necessary, returned (ibid., 193). *Balanced reciprocity* corresponds to direct exchange, where the given and the received is closely matched (ibid., 194). Reciprocation may be immediate or delayed, but usually a gift must be responded to within a relatively short term.

In his comparison between generalized and balanced reciprocity, Sahlins makes an interesting point (ibid., 195): 'It is notable of the main run of generalized reciprocities that the material flow is sustained by prevailing social relations, whereas, for the main run of balanced exchange, social relations hinge on the material flow'. *Negative reciprocity* is the attempt to get something for nothing, and it includes barter/trade (profit), theft, and warfare (conquest) (ibid., 195). In contrast to the other two forms of reciprocity, it is impersonal, and the participants represent opposed interests. As explained by Sahlins in diagrammatical form (Figure 23), the specific form of reciprocity is defined by kinship distance.

The commodities circulated in a tribal exchange system may be functional (consumables, such as raw materials and everyday objects) as well as symbolic items. Symbolic items may be either particularly well-executed/over-sized artefacts, or they may be functional pieces, or raw materials, which gained in symbolic value due to the distances they travelled, or both. As stated by Beck & Shennan (1991, 138), '... the spatially distant, because of its strangeness,

has great power'. On occasion, exotic objects may be fairly inconspicuous artefacts, but by travelling far, an exchange object is automatically transformed from being mundane to being special (Gould 1980, 142). Most exotic objects (for example raw materials) would be easily recognisable as such, due to their 'strange' aesthetically pleasing appearance, and frequently these items (in particular raw materials) had ancestral resonance (eg, Topping 2005, Appendix 1).

Due to the fact that exchanged symbolic/exotic items gave the recipient more status than exchanged consumables, the different types of goods tended to be exchanged in different ways. Consumables were mostly exchanged in relatively informal ways, whereas objects of higher symbolic (and thus status-giving) value might be exchanged in a more ceremonial and cyclic manner. At these events, kinship relations would be 'confirmed' and alliances forged, thus re-affirming or expanding the involved Big Men's local and regional positions. Stewart (1994, 90) suggests that the exchange in exotics and other symbolically laden objects kept the archaeologically invisible necessities moving through the system, and Earle & Ericson (1977b, 10) claim that the ceremonial cycles, where status-giving objects were exchanged and, on occasion, sacrificed/consumed, were the social mechanism which regularized and stabilized regional exchange and thus underpinned social renewal and stability.

Many exchange items, particularly consumables, did not travel far, but in some cases it is possible to follow objects over vast distances. They were distributed through extensive networks, either by means of down-the-line exchange (which could be ceremonial as well as non-ceremonial) or via regional transaction centres (directional exchange), which frequently functioned as combined focal points of economic, political, and ritual activities (Struever & Houart 1972, 52, 61). The aboriginal exchange network of Australian Arnhem Land is an example of a ceremonial exchange cycle based on down-the-line transactions (Orme 1981, 186), whereas the aboriginal interlocking networks of the American Pacific Plateau and Middle Missouri are examples of more centralized exchange networks (ibid., 189).

The Hopewell Interaction Sphere is an archaeological example of an exchange system, which was centred on Ohio and Illinois in the United States, but which also included more distant parts of the North American continent. Initially, the phenomenon was perceived as regional developments of a unitary 'Hopewell Culture', but later intense investigation of the 'Hopewellian' indicates that this material expression instead represents different culture types bound together by regular interaction (Struever & Houart 1972; Yerkes 2002). This exchange of material goods and ideas across different regional traditions was defined as an interaction sphere (Caldwell 1964), where many traditional exchange systems tend be confined to more closely related groups. Another important aspect of the Hopewell Interaction Sphere is the local reinterpretation of diagnostic Hopewell artefact and monument forms and ideological concepts.

The Hopewell Interaction Sphere was organized as a hierarchy of networks, where some goods circulated locally and regionally, and others interregionally. Struever & Houart (1972, 78) suggests that the local and regional exchange networks *within* the interaction sphere are characterized mainly by the exchange of subsistence-related artefacts, whereas the interregional interaction sphere is bound together by the exchange of status-related objects (also Struever & Houart 1964, 88). All levels of exchange are associated with transaction centres, characterized by complex ritual and burial earthworks. In relation to the discussion of the exchange in Arran pitchstone (Chapter 7.5), it is relevant that a substantial proportion of the Hopewell exchange goods was made up of exotic material, partly in the form of finished objects, but also in the form of unworked raw materials (Struever & Houart 1972, 68).

7.5 Discussion - the Distribution of Archaeological Pitchstone

7.5.1 The Distribution

In this chapter, the distribution of archaeological pitchstone in northern Britain is illustrated in a number of ways. It is hoped that the various forms of distribution maps and diagrams, in conjunction with the evidence presented throughout this volume, may allow a meaningful discussion of the distribution of archaeological pitchstone and its underlying tribal reality – the territorial structure of northern Britain and the exchange network responsible for the dispersal of Arran pitchstone.

During the production of the project's pitchstone database, a number of distribution patterns became increasingly obvious, and a number of preliminary zonated maps were presented as parts of papers and notes on the project's progress (Ballin 2006a; 2006b; 2007a; 2008a). Initially, this zonation embraced three zones, later four, and finally these were further sub-divided (Figure 4). The zonation displayed in Figure 4 was then used as a basis for the presentation of pitchstone finds across the various regions of northern Britain (Chapter 4.2). As this geographical division is relevant to the following discussion (prehistoric territories and exchange), the pitchstone zonation is briefly re-capitulated below.

In general terms, the zones are defined by a combination of decreasing assemblage size and decreasing frequency of pitchstone-bearing sites with increasing distance to the raw material outcrops on the Isle of Arran. Initially, it was thought that the zones were also defined by a decreasing frequency of pitchstone implements with increasing distance to source (eg, Warren 2006, 36), but this has been shown not to be the case (Table 5; Figure 4). In fact, the tool ratio increases with distance, from *c.* 10% in Zones I and II to *c.* 25% in Zones III and IV.

As shown in Table 3, there are considerable differences between the find frequencies in the various zones: Zone I (Arran) stands out with an average number of pitchstone artefacts per site per zone of 230; in Zone IIW this frequency is 14; in Zone IISW, it is 33; in Zone III, three; and in Zone IV, it is two. Orkney, in the Northern Isles, stands out with a find frequency of 14. As explained in Chapter 4.2.3, it is thought that this picture reflects prehistoric reality, but it is also influenced by sampling bias (Chapter 7.2). Since the writing of Chapter 4.2.3, fifteen pitchstone artefacts have been recovered from fields around Rothbury in Northumberland (K.L.R. Pedersen pers. comm.), and the general distribution map Figure 24 has been produced. The finds from Rothbury suggests that Zone III may include parts of northern (or maybe only north-east) England.

On the basis of the finds, as presented in the chapters above, it is now possible to define the four pitchstone zones, and their sub-zones, in the following summary form (Table 19):

The various trends displayed in Table 19 present information on chronology as well as prehistoric territorial structures and exchange. This information is discussed in greater detail below (Chapters 7.5.2-3).

The raw distribution of archaeological pitchstone is shown as Figure 24. It was decided to display the distribution in as simple a format as possible and, in this map, find spots were first plotted as dots per location. However, this manner of

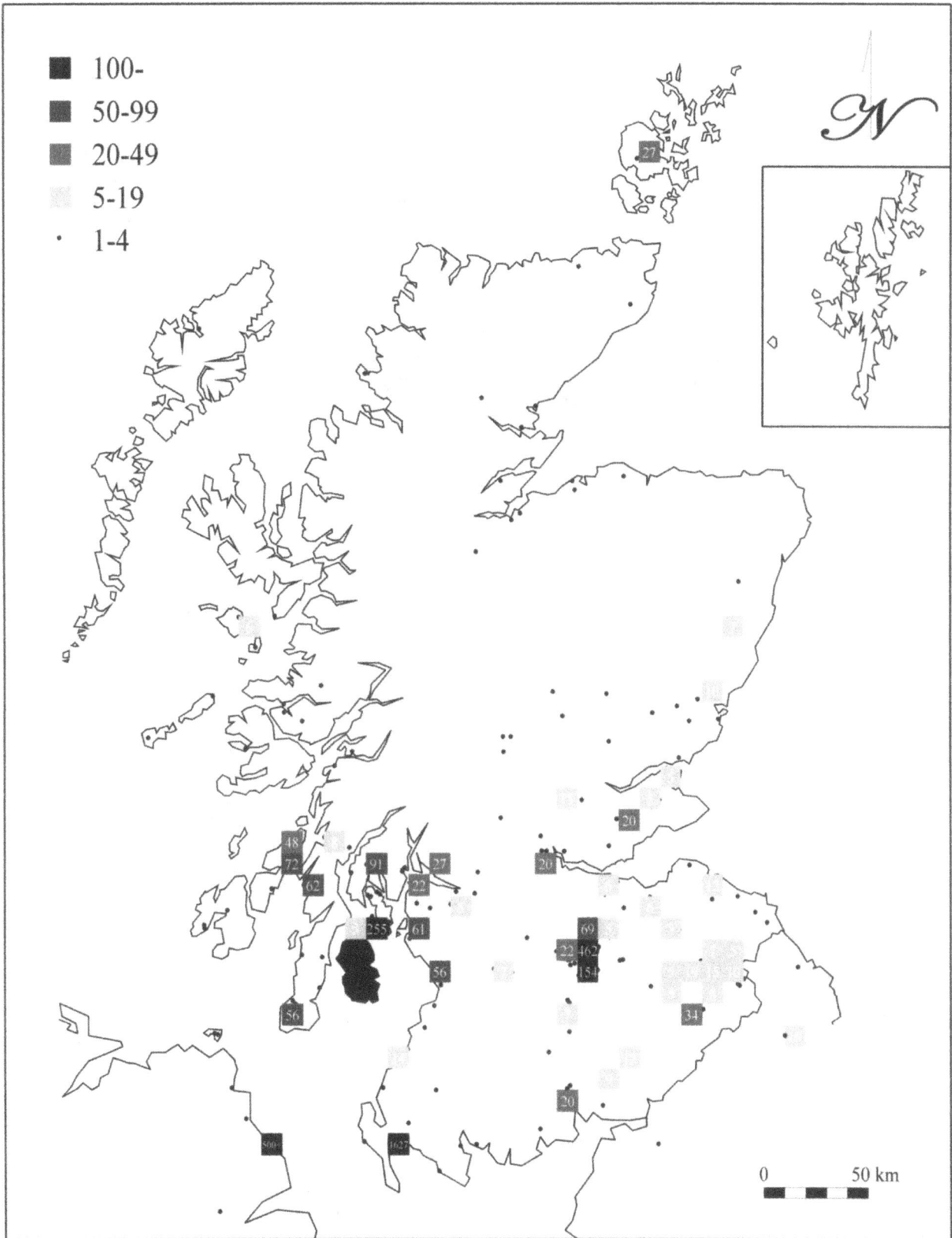

Figure 24. Distribution of all archaeological pitchstone in northern Britain.

Zones	Av. nos	Largest no. per 10x10 km	Porph. ratio	Microblade ratio	Levall. techn.	Inclusion in EBA burials	Tool ratio
I	>200	9363	C. 45-50	<15	X	X	13
IIW	C. 15-30	1627			(X)	X	C. 5-7
IISW							
III	C. 2-4	34	C. 0-3	C. 20-30	Absent	Absent	C. 20-30
IV		7					
Orkney	14	23	31	0	X		

Table 19. Summary characteristics of the various pitchstone zones.

display on its own would have presented a skewed picture, as at some locations few finds were recovered, and at others many were recovered; also, in several cases, dots of closely situated find locations covered each other. The map was therefore altered to show the distribution of archaeological pitchstone as a combination of dots and differently shaded 10 x 10 km grid units, where grid units with the most finds per area are darker than others.

The immediate impression of the pitchstone distribution is that there is a relatively dense concentration of rich find locations in the immediate vicinity of, and mainly north of, Arran (up to 70 km from the main outcrops of the Corriegills district). Towards the south and east, there is a relatively wide, find-poor zone, with a ring of large concentrations farther away from the sources (at a distance of approximately 80-100 km). Most likely, there are further relatively large concentrations at slightly greater distances from the Arran sources, such as in Zone III, and in northern England (eg, Northumberland), but this hypothesis needs to be tested by new research. Apart from on Orkney, pitchstone-bearing sites north of Argyll & Bute and Angus (ie, in the Highland region and in the north-east) generally only yield one piece per site.

To aid the discussion of the territorial structures of Neolithic northern Britain, Figure 25 was produced. It is a map of central and southern Scotland, where Thiessen polygons have been drawn around the most find-rich clusters of locations. It was only possible to do this for central and southern Scotland, as it is only here that there are sufficient numbers of finds of archaeological pitchstone to allow the production of Thiessen polygons. As there are many closely situated and roughly equally sized clusters in the area north and west of Arran it was decided not to subdivide this area further, as the resultant Thiessen polygons would be unrealistically small compared to those of the rest of central and southern Scotland. Although it must be admitted, that there are probably many more pitchstone-bearing locations to be found, and that the recovery of additional pitchstone artefacts may affect the produced Thiessen polygons somewhat, the fact that the polygons east of Arran are of approximately the same general size suggests that they may reflect some degree of prehistoric territorial reality.

It was thought that the production of fall-off curves might

be potentially fruitful to the discussion of the exchange network responsible for the dispersal of archaeological pitchstone throughout northern Britain. As Figures 24 and 25 suggest that two different distribution patterns may have existed, one involving the area north of Arran, and one probably the remainder of northern Britain, it was decided to produce two fall-off curves (Figures 26-27) instead of just one. If only one fall-off curve had been produced, the two different patterns would probably have cancelled each other out.

In both cases, the highest peak has been artificially lowered to allow minor peaks to become visible. Figure 26 shows how the distribution north of Arran may have been affected, or even determined, by the area's archipelago/fiord character. The two main peaks represent finds from sites on Kintyre and Bute, and in southern Argyll, and the many low peaks represent finds from smaller islands and fiords along the west-coast. Figure 27 shows the fall-off curve for the distribution in the remainder of northern Britain, with three noticeable peaks at roughly equal distances to each other. The first relatively small peak represents finds from locations along the Scottish mainland's west coast, where the exchanged pitchstone is thought to have 'made landfall'; the second and largest peak represents the accumulated finds from the three massive pitchstone collections of Ballygalley, Co. Antrim (in excess of 500 pieces), Luce Bay in Dumfries & Galloway (*c.* 1700 pieces) and Biggar in South Lanarkshire (*c.* 700 pieces); and the third peak represents the finds from the Tweed valley and Stirling/Fife.

7.5.2 Territories

The distribution maps in Chapter 7.5.1, in conjunction with other evidence presented in previous chapters, indicate a territorial structure for Neolithic northern Britain which includes several layers of social territories. As the subject of this volume is the archaeological pitchstone of northern Britain and, particularly, its distribution, the territorial structure suggested below is the structure indicated by the pitchstone exchange. This means that, in the present context, it is possible to say more about the territories in the vicinity of Arran, as well as the relations between these territories, than it is to discuss territories further away from the island, simply as a consequence of the varying volumes of evidence (ie, archaeological pitchstone).

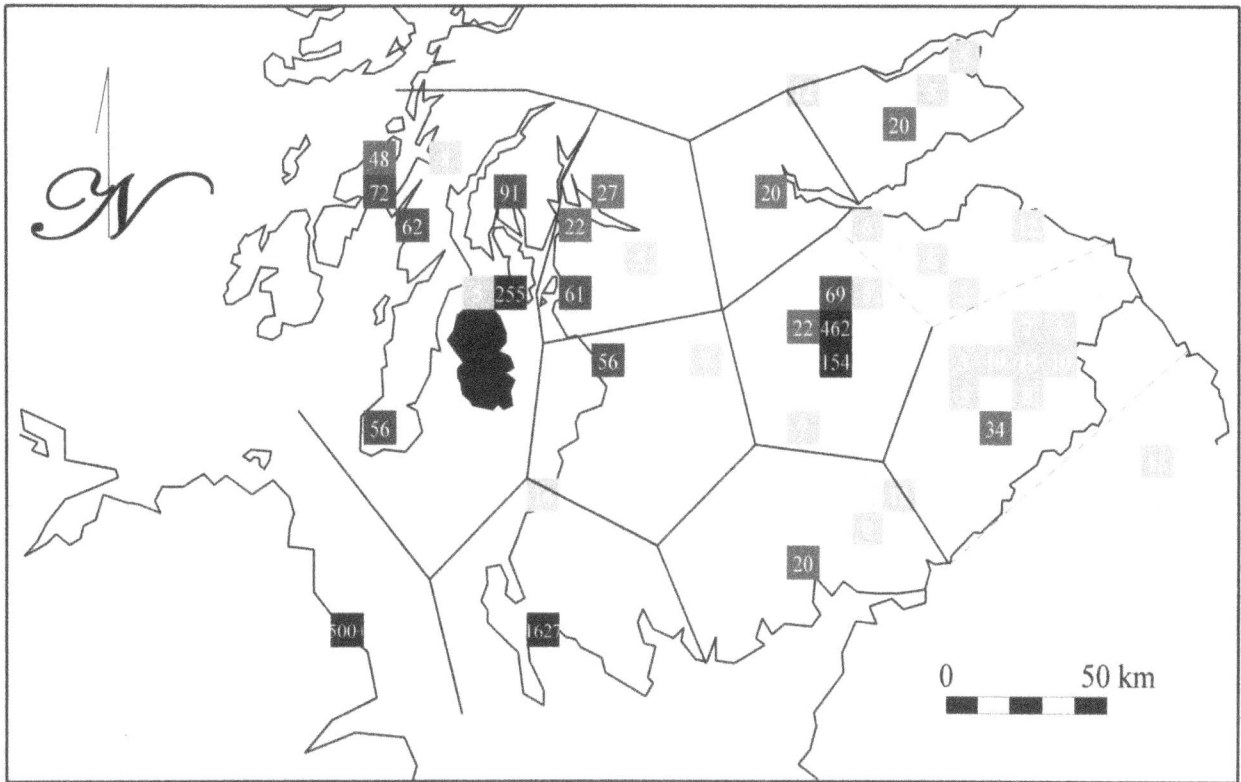

Figure 25. Sub-division of central and southern Scotland by Thiessen polygons.

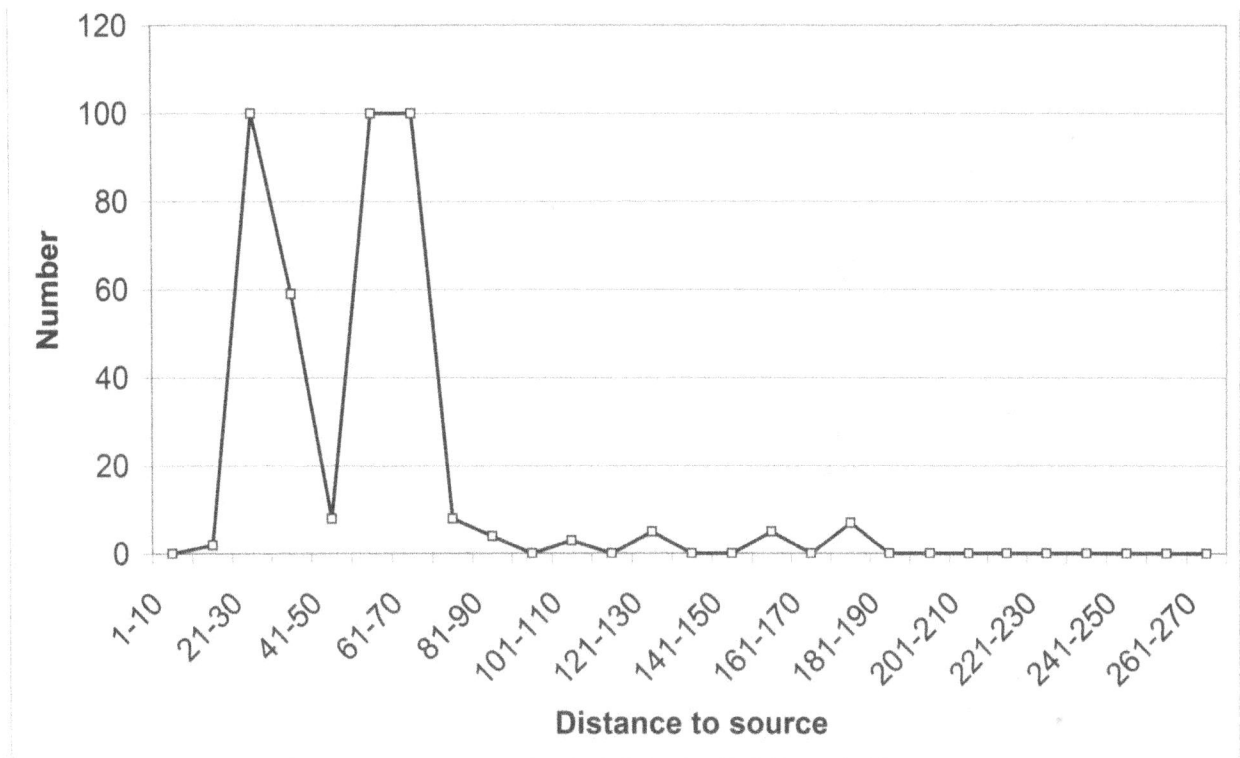

Figure 26. Fall-off curve for the area north of Arran.

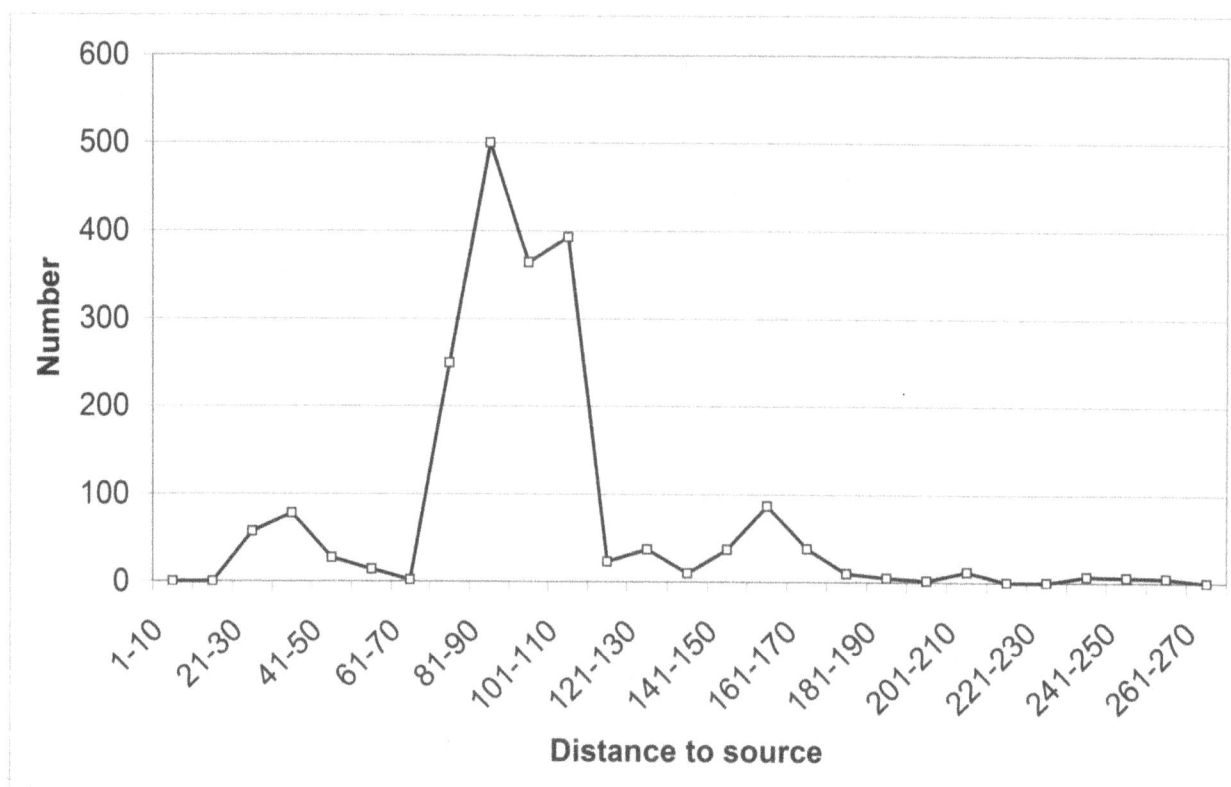

Figure 27. Fall-off curve for the remainder of northern Britain.

As indicated by Figures 24 and 25, two different territorial structures and associated exchange networks may be represented in northern Britain: one covering western Scotland (Zone IIW), and one the remainder of northern Britain. Most probably, the topographical realities of these two areas created two distinctly different socio-economical models, where the former is characterized by a multitude of spatially 'tight' microcosms in the form of Norwegian-style fiords and archipelagos, and the other by large, open tracts of land, for example on the mainland east of Arran. Consequently, the former developed smaller territories and included more local centres, and the latter larger territories and fewer local centres. This scenario brings the Marxist slogan to mind, that '... the material basis determines the ideological (social, political, religious, etc.) super-structure' and, without entering further into Marxist economy, this is probably the case here.

Table 19 gives an overview of the general territorial structure, and its various levels and forms of integration, from the perspective of Arran:

Level 1

Arran (Zone I) has a special position as the source of probably all archaeological pitchstone found in northern Britain (Williams Thorpe & Thorpe 1984; Simpson & Meighan 1999; Preston *et al.* 1998). It has the highest average number of pitchstone artefacts per pitchstone-yielding site (>200), and the location producing the highest

number of pitchstone artefacts (9,363) was found here (Site CTSF; GUARD's Arran Ring Main Water Pipeline Project; Donnelly & Finlay forthcoming). On Arran, pitchstone was exploited from the Late Mesolithic period through to, and including, at least the Early Bronze Age. Technologically, Arran was characterized by three main industries: a narrow-blade industry (LM-EN), a broad-blade/Levallois-like industry (LN), and a plain flake industry (EBA). As production of broad blades and flakes increased, in the middle and later Neolithic, the exploitation of porphyritic pitchstone sources increased. Most likely, Arran was in the control of one social group, for example a specific clan, and the fact that pitchstone dominates the island's lithic assemblages completely (commonly 60% to 100%, depending on period), suggests that this raw material may have represented what Wiessner (1983, 257) referred to as emblemic style ('... formal variation in material culture that has a distinct referent and transmits a clear message to a defined target population about conscious affiliation or identity'). No matter whether pitchstone had totemic or other specific meanings to people on Arran, they most likely used this raw material to demonstrate group identity (the 'People of the Pitchstone Island').

Level 2

The evidence suggests that, although Arran and its pitchstone sources may have been in the control of one local group, Arran probably formed part of a larger territory with, or was closely allied to, Zone IIW (ie, Argyll & Bute). This scenario is suggested by the following facts: 1) If Blackpark Plantation East on Bute (Ballin *et al.* forthcoming) was fully excavated, this location would probably be the individual site outside Arran yielding the highest number of pitchstone artefacts (presently, fieldwalking and test excavation has provided close to 500 pitchstone artefacts); 2) Zone IIW is characterized by a dense concentration of sites with high numbers of pitchstone artefacts (Table 20); 3) although Mesolithic pitchstone artefacts have not been identified, the discovery of pitchstone flakes in one of the Monybachach cists (Scott 1998) indicates that volcanic glass was used in Zone IIW as long as it was used on Arran, that is, well into the Early Bronze Age; 4) the frequency of broad blades in pitchstone and the ratio of porphyritic pitchstone are higher in this zone than in Zones IISW, III and IV. In Zone IIW, the ratio of pitchstone is not high enough to have made pitchstone a stylistic element in the same manner as on Arran but, obviously, the higher the local pitchstone ratio, the more a local Big Man was able to demonstrate his links to a powerful, usually kinship-related, ally. Arran and Zone IIW also share common pottery forms (the Rothesay/ Achnacree tradition; Scott 1977).

Level 3

Zones IISW and Zone II Ireland obviously had close ties to Arran, as demonstrated by the high frequencies of pitchstone finds around central places, such as the Biggar area in South Lanarkshire, the Luce Bay area in Dumfries & Galloway, and Ballygalley in Northern Ireland (500+ to *c.* 1700 pieces). Between these centres, relatively few large pitchstone-bearing assemblages have been recovered giving a lower general density of large assemblages than in Zone IIW. Diagnostic attributes and find contexts suggest that, in this area, pitchstone use is an entirely Early Neolithic phenomenon (Chapter 5): the microblade ratio is considerably higher than in Zones I and IIW (where the finds include elements produced by Late Neolithic and Early Bronze Age broad-blade and flake industries) and, as a consequence (Chapter 6), practically no porphyritic pitchstone has been recovered from the zones of this level. The almost 1,700 pieces from the Luce Bay area include one piece in coarsely porphyritic pitchstone, and the *c.*

	Numbers
Blackpark Plantation East, Bute (Ballin *et al.* forthcoming)	C. 500
Auchategan, Glendaruel, Argyll (Ballin 2006)	90
Lussa Wood 1, Jura (Mercer 1980)	67
Ellary Boulder Cave, northern Kintyre (Tolan-Smith 2001)	62
Balloch Hill, southern Kintyre (Peltenburg 1982)	58
Lealt Bay, Jura (Mercer 1968)	34
Midross, Loch Lomond, Argyll (Ballin forthcoming d)	27

Table 20. Significant pitchstone assemblages in Argyll & Bute.

700 finds from the Biggar area also include one piece of coarsely porphyritic material. Where some assemblages with pitchstone on Arran and in Zone IIW have been associated with the Late Neolithic Levallois-like technique, no pitchstone blanks from Zone IISW or from Northern Ireland seem to have been produced in this technique. It has also not been possible to date artefacts to the Mesolithic or Early Bronze Age periods. As in Zone IIW, the high pitchstone frequency of certain sites and areas is indicative of regular links and probably kinship-relations/alliances between some local Big Men and Arran. Zone IISW and Arran/Zone IIW also share common burial traditions, such as monuments of the Clyde Group. At least parts of Zone IISW are distinguished by a local burial tradition, referred to as the Bargrennan Group (Kinnes 1985, 33).

Level 4

Zone III shares characteristics with Zones IISW and II Ireland, such as dating elements. In this area, pitchstone was probably used during the Early Neolithic period, as indicated by the high microblade ratio, but not in later periods, as indicated by 1) the absence of attributes associated with the Levallois-like technique and 2) a general lack of pitchstone-bearing Early Bronze Age features. Relatively large assemblages are present although the maximum numerical size per 10 x 10 km grid is reduced from almost 1,700 in Zone IISW to 34 in Zone III. In terms of integration – that is, communication/contact with the source island of Arran – Level 4 has a distinct border-line appearance as it is neither particularly central nor peripheral. Although relatively large assemblages may be found, the average assemblage size in Zone III and Zone IV is *c.* two to four, where it is *c.* 15-30 in Zones IIW/IISW and >200 in Zone I. The tool ratio is also higher than in the more 'central' zones (in relation to Arran), with a tool ratio in Zone III comparable to that of Zone IV and Orkney (*c.* 20-30, where the ratio of Zones I and IIW/IISW is *c.* 5-13). This latter phenomenon indicates how pitchstone increased in value as it became more and more exotic, and how at increasing distances from Arran more effort was invested in not wasting this precious resource.

Level 5

In relation to Arran, this level (Zone IV) is distinctly peripheral: The average number of pitchstone artefacts recovered from a site yielding pitchstone is exceedingly low (Zones III and IV: *c.* two to four); the highest number of pitchstone artefacts within one 10 x 10 km grid unit is equally low (seven pieces); and pitchstone-bearing sites are generally rare. The latter may, to a degree, represent sampling errors, in the sense that more local development and subsequent archaeological activity would provide considerably more sites, but it is a fact that all pitchstone-yielding sites which *are* recovered within Zone IV only include small numbers of specimens. As at Level 4, the tool ratio is high, indicating that pitchstone, due to its exotic status, was a highly valued resource. As in Zones IISW and III, all diagnostic elements indicate that pitchstone use in Zone IV is an exclusively Early Neolithic phenomenon.

Orkney

Orkney seems to represent a special case, and although this may partially be due to sampling errors (that is, a higher than average level of archaeological activity, as well as a sharper focus on special places), some facts indicate that this is not the whole picture. The high average number of pitchstone artefacts per pitchstone-yielding site (14 pieces), as well as the zone's high maximum number (23 pieces) could possibly be explained in this manner, but the fact remains that Orkney is presently the only part of northern Britain where pitchstone has been safely associated with a Late Neolithic site, apart from on Arran itself (eg, Machrie Moor; Haggarty 1991, 65) and in Argyll & Bute (eg, Blackpark Plantation East; Ballin *et al.* forthcoming). The Barnhouse settlement (Richards 2005) is associated not only with pitchstone artefacts produced by the application of Levallois-like technique, but also Grooved Ware pottery, as well as buildings traditionally associated with the Orcadian Late Neolithic (eg, Skara Brae; Clarke & Sharples 1985). As a late assemblage, the Barnhouse finds include several lightly porphyritic pieces and it has no microblades. As a peripheral zone, Orkney's tool ratio is as high as in Zones III and IV. The high pitchstone frequency, as well as the fact that the pitchstone-using period of Orkney apparently corresponds to that of Arran and Zone IIW, indicates that Orkney may have had closer links with Arran than most of the other zones between the centre of the pitchstone exchange and its periphery.

As Figure 25 shows, it is possible to subdivide central and southern Scotland into a number of roughly equally sized 'lower-level' territories, based on the distribution of the main concentrations of pitchstone (Figure 24). The relatively equal sizes of these areas suggest that this subdivision may have some prehistoric relevance, but there are still pitchstone-yielding sites to be discovered and excavated, and particularly the northern (Falkirk, Edinburgh, the Lothian counties) and southern (the Scottish Borders) halves of Zone III south of the Firth of Forth need attention. Recent development of the infrastructure around and in Edinburgh has resulted in the recovery of several pieces of pitchstone, but in terms of pitchstone quantities and distribution in general, the Lothian counties are probably as poorly covered as the Ayrshire counties used to be (Chapter 7.2). Very little development has taken place in the Scottish Borders, but relatively intense amateur activity has shown that this area is potentially very interesting, and the excavation of Early Neolithic sites in the Tweed Valley may prove revealing. As mentioned above, the distribution pattern in the area north of Arran is somewhat different from that experienced in the rest of central and southern Scotland, probably due to very different topographical realities.

The definition above of different levels and forms of integration, in conjunction with the subdivision of central and southern Scotland into smaller territorial units, indicates a multi-layered territorial hierarchy:

- There seems to be a main territorial division between Zone IV, on one hand, and the rest of northern Britain. In simplified terms this represents a division between highland Scotland (including the lower-lying areas of north-eastern Scotland), and lowland/upland Scotland (including Ireland). In the northern part of Scotland, pitchstone is rare, and finds are mostly limited to one, usually tiny, piece per pitchstone-yielding site, whereas in the remainder of northern Britain it is not uncommon to find several, or even numerous, pieces of pitchstone per site. In terms of prehistoric people's perception of pitchstone, the distribution is interpreted as probably representing a three-step scale, sliding from almost entirely functional to almost entirely symbolic. On Arran, where pitchstone was ubiquitous, this raw material may have been perceived very much in functional terms, although it almost certainly had stylistic (?totemic) connotations. In Zones II and III, pitchstone was probably still a highly functional resource, although its value would have increased with growing distance to the sources on Arran, and being a strange and exotic raw material it would have been perceived much more in a symbolic light than on Arran. In Zone IV, where pitchstone would have been exceptionally rare, and thereby exceptionally strange, its value would have been predominantly symbolic.
- Due to the few pitchstone finds recovered from northern Scotland, it is not possible to subdivide this territorial unit further (unless other find-groups are drawn in), but central and southern Scotland is easily subdivided into smaller territorial units. Although the initial zonation was highly subjective, and based on modern administrative units (counties), the zones probably correspond to medium-level territorial units. These units were explained above in connection with the discussion of varying levels of integration in this part of northern Britain.
- These territorial units, or zones, may then again be subdivided into units the areas of which correspond roughly to those of modern counties (Figure 25). Most probably, the different topography of the area north of Arran (a Norwegian-style fiord and archipelago landscape/seascape) determined how this area was subdivided into slightly smaller local units than the remainder of mainland northern Britain.

At present, it is difficult to place Orkney and Ireland in this structure but, as mentioned previously, Orkney is a special case, and may in some way have had close relations with the territorial unit of Arran/Zone IIW. Although Ireland may have had very much the same form of link with Arran as Zone IISW, the divide formed by the Irish Sea, and the fact that Ireland followed distinctly different traditions in terms of lithics, ceramics, and monument building in general (eg, Woodman *et al.* 2006; Herity & Eogan 1996), suggests that this area was a separate territorial unit at the intermediate territorial level.

7.5.3 The Exchange Network

In this chapter, it is attempted to present the pitchstone exchange network by following Plog's approach (with later adaptations; Table 16).

Analytical focal point

The territorial structure and the exchange network presented in this volume are based on Arran pitchstone, and the flow of this unique raw material from Arran and into the world around the island. Focus on the exchange in other contemporary raw materials, such as Cumbrian tuff (Bradley & Edmonds 1993) and Antrim flint (eg, Saville 1999), would probably emphasize other areas or territories, but most likely the same divisions and steps in the territorial structure and exchange system would have been defined. As Cumbrian tuff was exchanged across an area as extensive as Arran pitchstone, the territorial subdivisions of northern Britain suggested by the exchange in pitchstone could, and in the future should, be tested by carrying out a similar analysis of the flow of Cumbrian tuff into and through northern Britain.

Duration

As indicated in Chapter 5, the exchange in Arran pitchstone was probably largely an Early Neolithic phenomenon, with the exceptions of Arran, Zone IIW, and Orkney. It appears that, in the Mesolithic period, pitchstone was used locally on Arran, but not subjected to wider exchange. The exchange in pitchstone from Arran and across most of northern Britain started relatively abruptly immediately after the Mesolithic/Neolithic transition, and it clearly had its peak in the first half of the Early Neolithic period. This is reminiscent of the rhyolite exchange in south-west Norway, which started equally abruptly after the beginning of the Early Neolithic period (traditionally defined as 5200 BP), although this resource may have been delimited to one specific social territory, and used very much as a stylistic marker of identity throughout that territory and its regional exchange network (Ballin 2007b). In northern Britain, the exchange in Cumbrian tuff probably started at approximately the same time (as noted by Tam Ward in connection with his work in the Biggar area of South Lanarkshire; Ness & Ward 2001), and it very much appears as if the development of extensive exchange networks in Great Britain is associated with the socio-economical changes that took place around the Mesolithic/Neolithic transition (also see papers in Clough & Cummins 1979; 1988).

The pitchstone exchange continued into the Late Neolithic period on Orkney, and in Zone IIW probably into the Early Bronze Age period. As mentioned above, this may be due to closer links between these two areas and Arran, and it is possible that Zone IIW formed one larger territorial unit with Arran. In the remainder of northern Britain, the flow of pitchstone may have decreased in the second half of the Early Neolithic period, and the pitchstone exchange had probably more or less come to a complete halt by the onset of the Late Neolithic (with the period being defined by the use of Impressed Ware and Grooved Ware). Although two pieces of worked volcanic glass demonstrate that pitchstone still 'trickled' from Arran eastwards by the beginning of the Late Neolithic period (one chisel-shaped arrowhead from the Luce Bay area and one from the Biggar area), the total absence of pitchstone blanks in the area outside Arran/Zone IIW and Orkney, which had been produced by the application of the Levallois-like technique, indicates that bulk exchange had stopped and that only the occasional individual piece was now exchanged. Instead, a massive flow of Yorkshire flint now moved in the other direction, from the east towards the west, more or less saturating the lithic production systems of the Scottish Borders and South Lanarkshire. This phenomenon is presently being investigated in connection with another research project (Ballin forthcoming i).

Magnitude

As indicated in connection with the discussion of the territorial structure of northern Britain, the quantity of pitchstone moving through the system varied with the distance to Arran and with the level of integration. Although the numbers of pitchstone artefacts recovered from Zone IIW (the Argyll & Bute fiords and islands) may be somewhat influenced by a relatively low level of local development, the comparatively high density of sites with high numbers of pitchstone (Figure 24), in conjunction with the potentially very large numerical size of the Blackpark Plantation East assemblage from Bute, indicates that in this zone pitchstone may have been exchanged in volumes smaller than those circulating on Arran, but still in quite substantial volumes. In Zones IISW and II Ireland, the flow was somewhat lighter, but noticeable concentrations are known. In Zone III the flow eased further, and in Zone IV it was reduced to a trickle (see discussion of the various levels of integration, above).

Diversity

It is difficult to define which other commodities circulated alongside pitchstone, and it must be assumed that many of the goods were perishable. Only Cumbrian tuff appears to have been exchanged throughout northern Britain at the same time as the pitchstone exchange took place, that is, in the Early Neolithic period. At present, pitchstone has not been recovered from Cumbria, and the most southerly pitchstoned-bearing site in this region is Carlisle (Fell 1990), although it would have been obvious to expect it here. At Biggar in South Lanarkshire, large amounts of Arran pitchstone and Cumbrian tuff were found together, and one would have expected the quarriers of the Great Langdale 'axe factories' approximately 150 km south of Biggar to have known of the existence of the volcanic glass from the north-west. Although the absence of pitchstone in Cumbria may be a result of erroneous raw material classification (as one of the most important raw materials in this area is a very pitchstone-like form of black chert; Peter Cherry pers. comm.), this phenomenon is slightly odd, and if pitchstone is truly absent from the region, this almost certainly reflects a deliberate decision not to let it in ('taboo').

Around the Irish Sea, the situation is distinctly different from that just described, as Arran pitchstone flowed in one direction (more than 500 pieces were recovered from Ballygalley; Simpson & Meighan 1999), and Antrim flint flowed in the other direction as demonstrated by several hoards and individual artefacts (Saville 1999). It is almost impossible to accurately estimate the volume of the exchange in Antrim flint, as the flint procured from the shores of south-west Scotland is identical to that found in Northern Ireland, and only pieces with soft cortex or particularly large pieces are certain to have come from Antrim. The date of the Portpatrick and Auchenhoan flint hoards in south-west and western Scotland is uncertain, but they may date to the later Neolithic (Saville 1999, 107). However, large well-executed flint tools from megalithic graves on Arran (eg, Bryce 1902, Figs 11-13) are undoubtedly in Antrim flint, and fragments of polished flint axeheads recovered at for example Auchategan and Midross, both in Argyll & Bute (Ballin 2006a; forthcoming d), also reflect exchange between the western parts of Scotland and Northern Ireland in the Early Neolithic period. The style of two flint arrowheads from Auchategan (Ballin 2006a) is similar to that characterizing some points from Early Neolithic Ireland and from the Isle of Man (Woodman *et al.* 2006, Fig. 4.27; Darvill 1995, Fig. 19), and they may have been exchanged as finished tools.

Axeheads in porcellanite were also exchanged between Northern Ireland and Scotland in the Early Neolithic period. By 1988 (Ritchie & Scott 1988, 88), eighty Scottish axeheads had been identified as porcellanite (Group IX), only numerically surpassed by axeheads in Cumbrian tuff (186 pieces). The third most numerous axehead group in Scotland are axeheads in Perthshire hornfels (Group XXIV), represented by 30 pieces. A direct link between the exchange in Arran pitchstone and that in porcellanite axeheads may be indicated by the fact that three of seven pieces of worked pitchstone recovered outside Ballygalley derive from Rathlin Island (Conway 1994; Sheridan pers. comm.), one of the two main sources of porcellanite (Sheridan 1986; Cooney & Mandal 1998, 58).

Boundaries

The geographical limits of the total pitchstone exchange system are presently uncertain. At this moment in time, pitchstone has been recovered from sites throughout Scotland (although in decreasing frequencies with increasing distances; see above), and only Shetland represents a pitchstone-free zone. However, this may reflect the level and form of local development, rather than prehistoric reality, and focused excavation of Early Neolithic sites in this island group may lead to the discovery of worked volcanic glass. This is by no means certain, though, as the flow of goods from north to south was almost nil, indicating a low level of integration. Ritchie & Scott (1988, 88) only knew of one certain axehead in Northmaven felsite (Group XXII) outside Shetland, that of Pencaitland in East Lothian; Ritchie (1968, 130) mentions two felsite adzes held by the Hunterian Museum and Art Gallery (one from Aberdeenshire and one from the Scottish Borders), as well as a felsite Shetland knife found in Lanarkshire (now in the Kelvingrove Art Gallery and Museum).

Northern Ireland was clearly part of the exchange network, as indicated by the massive concentration of pitchstone at Ballygalley, with individual pieces having been found at five other locations in Antrim. The most southerly pitchstone find in Ireland is that from Lambay Island in County Dublin (Graeme Warren pers. comm.), approximately 275 km south of the sources on Arran. A single piece of worked pitchstone has been retrieved from a site at Ballachrink on the Isle of Man (McCartan & Johnson 1991), approximately 175 km from Arran.

A small number of finds south of the Anglo-Scottish border indicates that the exchange network responsible for the dispersal of Arran pitchstone included at least parts of England (Ballin 2008a). If one considers the fact that, in Scotland, archaeological pitchstone has been found at Barnhouse on Orkney (Richards 2005), 400 km north of the outcrops, it should be possible to recover the occasional piece of pitchstone as far south as Manchester. If one also considers the fact that the northwards dispersal of pitchstone probably stops due to the barrier created by the Atlantic Ocean, it is not inconceivable that worked pitchstone could have travelled as far south as the English Channel. As mentioned above, the lack of pitchstone in Cumbria is difficult to explain, as axeheads in Cumbrian tuff were exchanged into Scotland, but future examination of Early Neolithic assemblages from Cumbria may change this picture. The recent (2008) find of 15 pitchstone artefacts from fields near Rothbury in Northumberland (Kristian L.R. Pedersen, pers. comm.), suggest that focused search for 'English' pitchstone may generally prove fruitful.

Acquisition

As the exchange in Arran pitchstone is generally an Early Neolithic phenomenon, with some later exchange taking place between Arran, Zone IIW and Orkney, it is highly unlikely that this was obtained by non-Arran people in the form of embedded procurement, that is, as a by-product of, for example, hunting trips, or trips defined by other activities. The numbers of worked pitchstone recovered in the zones around Arran (Zones II-III), as well as the find contexts of archaeological pitchstone (eg, domestic, burial, and ritual contexts; see for example Chapter 5.3.4), suggest that volcanic glass was a treasured material, and it was most likely obtained in a more focused manner.

If the territorial structure suggested above is accepted as a realistic model, direct acquisition is equally unlikely, at least for the more peripheral territories, where groups would have had to cross the territories of several other groups to get to Arran, whereas it is not logistically unrealistic to imagine that groups around the Firth of Clyde visited Arran on occasion. However, the frequency fall-off with increasing distance primarily supports the general dispersion of archaeological pitchstone by indirect procurement (ie, exchange), as direct procurement, had this been the general form of acquisition, ought to have brought about a more equal distribution of pitchstone throughout northern Britain. Direct procurement is thought to have been the general form of acquisition in connection with the dispersal of some raw materials (dolerite, greenstone, rhyolite) throughout western Norway in early prehistory (Bruen Olsen & Alsaker 1984; Bergsvik & Bruen Olsen 2003), and in these cases the result was a fairly even distribution within the implicated social territories, where these raw materials were clearly stylistic markers of group identity.

The question is what form the indirect procurement of pitchstone took? Were goods (including pitchstone) exchanged small-scale, village by village, and thus slowly dispersed throughout northern Britain, or may larger volumes of goods have changed hands for example at regional 'trade fairs' organized by local Big Men? Simple down-the-line trade should have resulted in a gradually declining fall-off curve (corresponding to Figure 18), whereas the actual pitchstone fall-off curve for northern Britain outside Arran/Zone IIW clearly has three noticeable peaks at roughly equal distances to each other (Figure 27), indicating some form of directional exchange. As mentioned in Chapter 7.5.1, the first relatively small peak is likely to represent finds from locations along the Scottish mainland's west-coast, where the exchanged pitchstone 'made landfall'; the second and largest peak represents the accumulated finds from the three massive pitchstone collections of Ballygalley in Co. Antrim (*c.* 500 pieces), Luce Bay in Dumfries & Galloway (*c.* 1700 pieces) and Biggar in South Lanarkshire (*c.* 700 pieces); and the third peak represents the finds from the Tweed valley and Fife/Stirling.

This directional exchange could have been part of a redistribution system, or the concentrations of pitchstone may pinpoint the locations of prehistoric regional 'trade

fairs'. Some relatively recent North American exchange networks, such as the Pacific Plateau network (Orme 1981, 189), included regular 'trade fairs' where groups from distant corners of a region met and exchanged local, frequently specialized or exotic commodities. As redistribution *sensu stricto* is a phenomenon usually associated with chiefdoms this economic mechanism presents an unlikely explanation for the Early Neolithic concentrations seen at Luce Bay, Biggar and Ballygalley, even allowing for the fluctuations suggested by Fowles' social trajectories model (Figure 22).

A number of factors support the idea of pitchstone dispersal by means of 'trade fairs':

- The three main centres, Luce Bay, Biggar and Ballygalley, are located at strategically logical distances to each other, with Luce Bay possibly servicing the Scottish south-west, Biggar central and eastern Scotland, and Ballygalley northern and eastern Ireland.
- All three centres are situated in logistically sensible locations in relation to potential regional exchange: either in sheltered inlets by major coastal 'trade' routes (Luce Bay and Ballygalley) or in the narrow gap ('the Biggar Gap') between the Clyde and the Tweed, thus linking the eastern and western parts of southern Scotland (Cummings 2002; Ballin & Ward 2008).
- In the Hopewell Interaction Sphere (Struever & Houart 1972; Yerkes 2002) the central places where fairs took place were generally associated with concentrations of elaborate burial and ritual monuments. It is quite possible that for example Biggar Common (Johnson 1997) may be just such a location, as it includes multiple burial and ritual features (albeit less elaborate than their Hopewell counterparts). However, it should be borne in mind that the features examined at this site represent activities from the Late Mesolithic to the Early Bronze Age. Topographically, Biggar Common is a promontory situated approximately one kilometre from the Clyde.
- All three centres are associated with other groups of exchanged commodities. Particularly fragments of axeheads in Cumbrian tuff are common in the Luce Bay area (some were examined in connection with a visit to Dumfries Museum; also see Maynard 1993) and in the Biggar area (some were examined in connection with a visit to Biggar Museum; also see Ness & Ward 2001). The lithic and stone assemblage from Ballygalley still has not been published in its entirety, but the collection is known to include material from the Great Langdale 'axe factories' in Cumbria (Cooney & Mandal 1998, 114).

Most likely, these 'trade fairs' would have been organized by local Big Men, *not* to underpin a local dynasty as in chiefdoms, but to forge/maintain largely kinship-based alliances and secure continued peaceful relations in the region. The acquisition of pitchstone probably formed part of exchange partnerships based on balanced reciprocity (Figure 23).

Directionality and system shape

As explained at the beginning of the present 'Plog-based' breakdown of an exchange network (Plog 1977; Thatcher 2001; Table 16), this analysis is focused on the exchange in Arran pitchstone, and the directionality of this commodity from the Isle of Arran, in the Firth of Forth, from which point all archaeological pitchstone apparently emanates (Williams Thorpe & Thorpe 1984; Preston *et al.* 2002; Ballin & Faithfull forthcoming). Based on the presently available evidence, this direction appears to have been largely west-east across the central and southern parts of the Scottish mainland, although currents of the exchange system also moved towards the west (Ireland), north (northern parts of Scotland), and south (England).

As indicated above, the exchange in pitchstone outside Arran/Zone IIW/Orkney was largely an Early Neolithic phenomenon, with the bulk of the recovered pitchstone being datable to the first half of this period. In the second half of the Early Neolithic, the exchange in pitchstone slowed down, and by the beginning of the Late Neolithic (as defined by the use of Impressed Ware/Grooved Ware) the flow of pitchstone had almost stopped completely. Instead, Yorkshire flint was now moving in the opposite direction into Scotland, and apparently took over the role of pitchstone as the main exotic lithic raw material being exchanged throughout northern Britain. This suggests that by the onset of the Late Neolithic period, Yorkshire may have gained in importance at the expense of Arran.

In a sense, it is difficult to speak of one single exchange system, as many different commodities were exchanged within a given local area, but with the different types of goods having different *foci*, moving in different directions, and having different densities in local archaeological settings. In his discussion of 'world systems' and 'centre/periphery', Renfrew (1993, 7) suggested that it may prove fruitful to perceive '... the trading systems of Neolithic Europe [...] as interlinked local systems'. This fits the present situation quite well: although in the Scottish south-west Arran pitchstone, Cumbrian tuff, and Antrim flint must have been exchanged at the same events ('trade fairs'?), and possibly exchanged against each other, these three materials clearly represent three individual, but overlapping, systems: on the present evidence, the core area of the pitchstone exchange appears to have been western, central and southern Scotland; the core area of Cumbrian tuff north-west and north-east England, with noticeable quantities having been recovered from southern Scotland and central England; and Antrim flint was mainly exchanged in northern and eastern Ireland, western Scotland and parts of the areas around the Irish Sea (eg, the Isle of Man).

In addition, there were less widely exchanged lithic raw materials, supplying smaller local systems, such as Rhum bloodstone, Staffin baked mudstone and Lewisian mylonite. These systems appear to have covered areas of up to *c.* 100 km radius. Recent evidence indicates that the former two raw materials may have been used and exchanged throughout prehistory and not only during the earlier prehistoric period, as popular belief suggests (Ballin 2008c; 2008d). Some raw materials, like quartz and chert, may not have been subjected to exchange, as they were ubiquitous and widely available within their regions, and they are both considered relatively poor raw materials in terms of flaking properties.

Object value

The value of exchanged raw materials is difficult to assess, as it may have been influenced by a multitude of factors, the most basic of which are functionality and symbolic issues. The former includes, *inter alia*, which tools can be made from a specific raw material (ability to form sharp and/or durable edges), flaking properties, accessibility (physically and socially), and abundance/rarity/absence. From a functional point of view, it is important that it is possible to produce the implements needed in the tribe's daily life from a specific raw material; that it flakes in a controllable manner; that it is relatively easy/cheap to obtain; and that it is present in the amounts needed daily. Raw materials with high symbolic values frequently have some of the above characteristics, such as good flaking properties and the ability to form sharp/durable edges, but a distinct and/or pleasing appearance may increase the value of a raw material, not least if the colours, patterns or other physical attributes associates the raw material with, for example, tribal mythology or legendary, frequently totemic ancestors (eg, Topping 2005, Appendix). In such cases, the rarity of a raw material, or the fact that it comes from distant places, adds value to it (Beck & Shennan 1991, 138; Gould 1980, 142).

The value of pitchstone is indicated by a number of factors, such as find context, distribution, and tool ratio:

- Although most archaeological pitchstone may have been recovered from domestic sites, a substantial number of pieces have been retrieved from burials, timber halls, and ceremonial contexts (for example pits) (see overview in Chapter 5.3.4).
- Throughout northern Britain the distribution of pitchstone appears to be relatively even, in the sense that pitchstone is recovered from a relatively high proportion of all Early Neolithic sites (eg, Table 11), although with find frequencies varying between zones. This suggests that a concerted effort was made to acquire the raw material, and that the distribution is not random. Or to put it differently: if the dispersal of archaeological pitchstone had been a question of simple cost-benefit calculations and functionality, the fall-off

should have taken the shape of a gradually sloping curve (such as Figure 18), and pitchstone should not have been expected to appear in archaeological contexts in, for example, Caithness or on Orkney.
- The fact that the tool ratio actually increases with increasing distance, indicates that, furthest from the sources on Arran, pitchstone was considered a highly valuable commodity, and that attempts were made not to waste this resource.

It is very difficult to determine exactly what value pitchstone had to prehistoric people, and which mechanisms regulated access, but most likely its specific value was determined by a variety of issues. Firstly, a commodity (for example a raw material) must be appreciated for its functionality, its striking appearance, and/or its association with parts of tribal mythology; and secondly, distance – more or less automatically – adds a premium to the value as a consequence of the time/labour invested in acquiring it, combined with a less measurable extra value determined by rarity in itself (an added 'mysterious' aspect; Beck & Shennan 1991, 138). The effect of distance on the value of exchanged commodities is explained by Sahlins in Figure 28.

The pitchstone zonation and the frequencies/distribution of archaeological pitchstone within the various zones, suggest that this raw material may have been perceived in three different ways, that is, it had different forms of value): 1) in Zone I (on Arran), and to a degree in Zone IIW (for example at Blackpark Plantation East), entire assemblages are in pitchstone, or they may be heavily dominated by pitchstone; 2) in Zones II and III, most lithic assemblages are dominated by local materials, but with substantial complements of pitchstone artefacts; and 3) in Zone IV, most pitchstone-bearing assemblages only include a single piece of pitchstone. Simply put, this is a difference between full assemblages, tool kits, and single pieces of pitchstone.

A. Trading groups

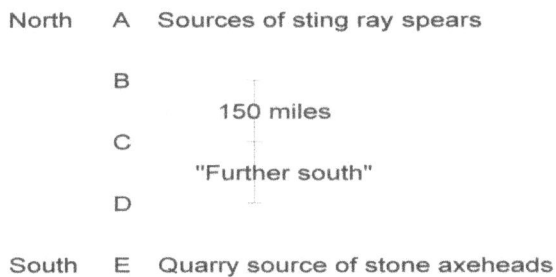

North	A	Sources of sting ray spears
	B	
		150 miles
	C	
		"Further south"
	D	
South	E	Quarry source of stone axeheads

B. Rates of exchange at various points

at B	12 spears = 1 axehead	
at C	1 spear = 1 axehead	
at D	1 spear = "several axeheads"	

Figure 28. An example of how the value of goods increases with increasing distance; the example is from the Queensland Trade Chain (after Sahlins 1972, Fig. 6.1; Sharp 1952).

The complete domination of pitchstone on Arran, and on some sites in Zone IIW, suggests that in this part of northern Britain – that is, amongst the people who controlled access to the pitchstone sources – pitchstone may have been viewed in a stylistic light, and possession of full assemblages of pitchstone may have been one of the ways these people indicated who they were and that they belonged together. It is highly likely that they would have associated this raw material with parts of their mythology, and that pitchstone may have been a link between them and one or more (totemic?) ancestors.

As the exchange network would largely have been based on kinship relations, anybody, for example on the Scottish mainland, who acquired pitchstone would automatically link into this mythology and at the same time indicate an alliance between them and people on Arran. It is highly likely that, with growing distances and growing pitchstone value, this raw material may increasingly have been used in more specialized parts of tribal life, for example in connection with the tribe's ceremonial life, but at present this is somewhat conjectural. It has been suggested that, among other things, pitchstone was used in connection with burials for '... drawing out the black visceral blood of internal organs' (Jones 1997), but this suggestion disregards the fact that pitchstone, on and off Arran, was modified into not only knives, but also arrowheads, scrapers, piercers, etc. The question of the specific value and use of pitchstone clearly requires more in-depth analysis of the find-contexts of archaeological pitchstone, and use-wear analysis of pitchstone from domestic, burial and ritual contexts may provide an at least partial answer.

It has occasionally been emphasized that pitchstone recovered off Arran tends to be high-quality aphyric material (eg, Warren 2006, 36). To some, this may suggest that pitchstone was exposed to a careful selection process prior to exchange, and that only the functionally and aesthetically better forms of pitchstone were desired off Arran. However, as indicated above (Chapters 5 and 6), the use of aphyric pitchstone is a general characteristic of Early Neolithic pitchstone-bearing assemblages throughout northern Britain, with porphyritic pitchstone becoming more common in connection with the later Neolithic broad-blade industries of Arran, Zone IIW and Orkney.

System complexity

As demonstrated above, the system responsible for the dispersal of archaeological pitchstone throughout the defined zones of northern Britain (and beyond?) was a complex one: the amounts of pitchstone flowing through the different zones varied distinctly; pitchstone is likely to have been distributed differently within the different zones; most likely, pitchstone was perceived differently in the various zones; and, consequently, it may have been used and disposed of differently.

As the Hopewell Interaction Sphere of the American Mid-West (eg, Caldwell 1964; Struever & Houart 1964; 1972; Struever 1972; Seeman 1995; Yerkes 2002) is one of the most thoroughly discussed and published tribal exchange systems, and as it – like the exchange system of northern Britain – seems to have had the dispersal of exotic raw materials as one of its key activities, it was chosen as the most appropriate comparative model by which greater understanding of the northern British exchange system could be achieved.

As discussed above, it was difficult to define precisely the geographical borders of a northern British exchange network, and it was equally difficult to list which goods other than pitchstone circulated within this network. Within, for example, the central part of southern Scotland, a number of exotic raw materials were exchanged side-by-side, but representing different networks (as seen from the perspective of the individual commodities), centred around different procurement centres: pitchstone emanated from the Isle of Arran, Cumbrian tuff from the Great Langdale 'axe factories', and Antrim flint from the shores of Northern Ireland. Beyond these zones, there are further overlapping networks, eventually linking the British Isles as a whole into one system, and, subsequently, linking the British Isles with Continental Europe, for example via the exchange of jadeite axeheads from southern France (Petrequin *et al.* 2002). This has been referred to as a 'world system' (eg, Renfrew 1993), and this system is obviously too extensive to be meaningful in the present context. A focused perspective is needed, and in this chapter focus has been on Arran pitchstone. It is suggested, with reference to the Hopewell (as the Hopewell Interaction Sphere is commonly abbreviated), to refer to the involved exchange system as the Arran Pitchstone Interaction Sphere, which may cover northern Britain, or it may eventually prove to cover the British Isles as a whole. This interaction sphere is defined as the area within which social groups interacted, directly or indirectly, with the 'Isle of Pitchstone', Arran.

In terms of defining the Hopewell Interaction Sphere, emphasis has been put on different elements. Struever (1972, 315), for example, writes that:

> '... the general word interaction [...] attempts only to indicate that some form (or forms) of communication, intercourse, or articulation existed prehistorically to enable far-distant groups to share an assemblage of imported raw materials, artefact styles, and precepts governing the interment of certain dead'.

He underlines (ibid., 303) that '... primarily raw materials and stylistic concepts, not finished goods, were moving through the network'.

In its focus on the exchange in exotic raw materials, the Arran Pitchstone Interaction Sphere corresponds to the Hopewell system, and northern Britain is clearly defined by a common material culture, as far as lithic artefacts are concerned. However, in contrast to the Hopewell Interaction Sphere which is generally characterized by the erection of complex burial mounds (eg, Seeman 1995), the Arran-centred system includes a higher degree of variation as associated pottery and mortuary practices vary distinctly from zone to zone, for example between the megalithic west/north and the non-megalithic east (eg, Kinnes 1985, Illus 6). This, however, does not disqualify the northern British system as an interaction sphere. In their paper, Struever & Houart (1972, 48) stress that the Hopewell Interaction Sphere represented '... different *culture types* between which some form of interaction existed [the author's emphasis]', and Seeman (1995, 123) points out that the intense interaction within the Hopewell did not necessarily require an increase in shared ideology.

The best way to explain some of the differences between the Hopewell Interaction Sphere and the Arran Pitchstone Interaction Sphere is probably to see the two systems as the interaction spheres of societies at different social levels of development: as demonstrated by Yerkes (2002, Table 1), the Hopewell includes mainly tribal elements but also some elements usually associated with chiefdoms (such as huge complexes of elaborate ritual and burial monuments), and the general consensus regarding the Hopewell is that the involved groups represent budding chiefdoms, although they are still largely tribal (eg, Caldwell 1964; Struever & Houart 1964; 1972; Struever 1972; Seeman 1995; Yerkes 2002). In contrast, the groups involved in the Arran Pitchstone Interaction Sphere are groups which just recently evolved from band level organization to tribal organization, and even allowing for the possible fluctuations indicated by Fowles' (2002, 20) social trajectories model, the social organization of Early Neolithic northern Britain was probably distinctly tribal.

In that respect, the centres indicated by one of the pitchstone fall-off curves (Figure 27) are more likely to represent, for example, 'trade' fairs organized by local Big Men rather than chiefly redistribution centres, and there are no signs that the exchanged pitchstone was used to underpin an upcoming or established elite. The dominating burial practice was still communal interment, and pitchstone was generally not disposed of in the form of sacrificial caches. Occasionally, individual pieces, or a handful, of pitchstone artefacts were deposited in ritual pits, such as at Fordhouse Barrow in Angus (Ballin forthcoming f), and only the deposition at Torrs Warren in Dumfries & Galloway (Cowie 1996) differs from the general pattern. Apparently, at Torrs Warren, a relatively large amount of pitchstone was destroyed by fire (cf. Larsson 2004), possibly as a sacrifice of a precious exotic resource. Otherwise, pitchstone seems to have circulated relatively freely in the Early Neolithic society as suggested by the recovery of much pitchstone from domestic locations, such as most of the Biggar sites (Ballin & Ward 2008).

8. Future Perspectives

Although the Scottish Archaeological Pitchstone Project has allowed the compilation of a substantial database of pitchstone artefacts from northern Britain, as well as a discussion of the mechanisms responsible for the dispersal of this material, the present volume – like the works of Mann (1918), Ritchie (1968), and Williams Thorpe & Thorpe (1984) – only represents another step on a 'ladder of inference' (Hawkes 1954). Since the publication of Williams Thorpe & Thorpe (1984), the distribution of archaeological pitchstone across northern Britain has become denser, but the distribution map is still incomplete. This has consequences for our understanding of the role of Arran pitchstone in specific parts of northern Britain, as well as for our insight into the size and character of the total Arran Pitchstone Interaction Sphere.

Some of the shortcomings of the present distribution maps (Figures 24 and 25) will be dealt with slowly over time, as new local development and associated archaeological fieldwork 'fill in the blanks', but other shortcomings ought to be dealt with via focused archaeological projects, involving local professional archaeologists as well as lay enthusiasts. In terms of geographical coverage, some of the more pressing issues are:

- The 'English Question';
- The status of Zone IIW;
- The status of Zone III;
- The status of Orkney;
- Northern Ireland/Ireland.

The question of the potential extension of the Arran Pitchstone Interaction Sphere into England, and how far it may have reached into that area, was discussed extensively above (also see Ballin 2008a). Basically, archaeological pitchstone should be expected from sites well inside England, and particularly the apparent absence of pitchstone artefacts in Cumbria is difficult to understand, as this region supplied large tracts of Scotland with Group VI axeheads at the peak of the pitchstone exchange. Presently, pitchstone samples are being distributed amongst archaeologists and enthusiasts south of the Anglo-Scottish border, and new finds of pitchstone have already been reported from Northumberland (Kristian L.R. Pedersen pers. comm.). It is hoped that new finds of archaeological pitchstone will also be made in other parts of England.

Based on a combination of the general density of pitchstone-bearing sites, the density of multi-piece sites, the general raw material and typo-technological composition of assemblages, and the dates of assemblages, it was suggested above that Zone IIW may possibly have formed part of a larger west-of-Scotland territory which, due to topographical factors, was organized differently from the territories of mainland Scotland and Ireland. This likely state of affairs may be tested by more detailed analyses of the available finds from this region, but it is more likely that fruitful tests would rely on new finds, such as more safely contexted assemblages from, for example, Late Neolithic and Early Bronze Age sites.

The status of Zone III is clearly somewhat uncertain, and the finds suggest a possible difference between the areas north and south of the Firth of Forth. Apparently, the density of pitchstone-bearing sites north of the Forth is lower than that of the Scottish Borders, but the two areas have been explored in so different ways (mainly excavations in the north and mainly fieldwalking in the south), that one is hardly comparing like with like. Basically, focused fieldwalking should be carried out in the counties north of the Forth, whereas excavations in pitchstone-yielding fields in the Scottish Borders may reveal concentrations comparable to, although probably not quite as numerically large as, those of the Biggar area in neighbouring South Lanarkshire.

It is difficult to test the status of Orkney and its possible close links to Arran/Zone IIW, but any test case should obviously focus on the situation in Caithness, and whether pitchstone forms part of Early and/or Late Neolithic assemblages in that area: is the apparent difference between Orkney and northern Scotland real or is it an artefact of skewed research? In Northern Ireland, two questions present themselves, namely 1) whether more large pitchstone assemblages may be found in the immediate vicinity of Ballygalley, and 2) whether pitchstone artefacts may be more common in Northern Ireland and in the Irish Republic than the present distribution map leads one to believe? The distribution of large pitchstone concentrations around Biggar in South Lanarkshire, and around Luce Bay in Dumfries & Galloway, suggests that more multi-piece pitchstone assemblages should be expected in the vicinity of Ballygalley, and it is possible that, in the Irish Republic, some pitchstone artefacts have been erroneously classified (for example as black chert; cf., Mann 1918, 44). The former question could be tested by focused fieldwalking, and the latter by re-examination of Early Neolithic sites from Ireland.

In this volume, the suggested territorial structure is undeniably sketchy. The borders of the defined 'upper level' territories, or zones (Figure 4), are influenced by modern administrative boundaries, for example between counties, and the borders of the suggested 'lower level' territories, as defined by the use of Thiessen polygons, are understandably of an artificial nature. The borders of the larger territorial units could possibly be tested by comparison with borders suggested by the distribution of local pottery styles and monument types, and the borders of both levels of territories ought to be adjusted by taking topographical features into account, such as rivers, mountain ranges, hill tops, promontories, etc. (O'Shea & Milner 2002, 208).

One of the most interesting issues relating to the exchange in Arran pitchstone is the chronological question: exactly when does systematic exchange in pitchstone start and end, and what are the mechanisms driving this development? As discussed in Chapter 5, there is no evidence to support the exchange in pitchstone beyond Arran in the Mesolithic period, and it seems that the start of the pitchstone exchange coincides rather precisely with the onset of the Early Neolithic period. This corresponds to the situation in south-west Norway, where the exchange in rhyolite from the Bømlo quarries started at the very beginning of the Early Neolithic period (Alsaker 1987). Although some small-scale exchange in lithic and stone raw materials may

have taken place in the British Mesolithic, it seems as if, in this part of the world, systematic procurement and exchange in these raw materials are inextricably linked to the socio-economical 'revolution' that took place at, and defined, the transition between the Mesolithic and the Neolithic periods, with changes to the basic economy, kinship structure, ideology, etc. This was discussed in Chapter 7.

The collapse of the pitchstone exchange is a less certain matter. Apparently, the exchange in Arran pitchstone peaked in the first half of the Early Neolithic period, and although it continued through the second half of that period, *systematic* exchange in this raw material had probably all but ceased by the beginning of the Late Neolithic. It is tempting to relate this process to the growing exchange in Yorkshire flint around the transition between the Early and Late Neolithic periods, but at present this phenomenon needs further scrutiny: 1) Did the pitchstone exchange just 'fizzle out' or did it come to an abrupt halt? 2) Exactly when did the exchange in Yorkshire flint begin – slowly through the later part of the Early Neolithic period, or more abruptly at the beginning of the Late Neolithic period proper? It is the author's working hypothesis that the change in exchange patterns, from Early Neolithic west-east orientated exchange to Late Neolithic east-west orientated exchange, is somehow associated with the socio-economical changes that took place during the transition. It is hoped that an ongoing project at National Museums Scotland (focusing on assemblages rich in Yorkshire flint from the Scottish Borders; Ballin forthcoming i) may shed some light on the altering exchange patterns in the later part of the Neolithic of northern Britain.

BIBLIOGRAPHY

Acquafredda, P., & Muntoni, I.M. 2008: Obsidian from Pulo di Molfetta (Bari, Southern Italy): provenance from Lipari and first recognition of a Neolithic sample from Monte Arci (Sardinia). *Journal of Archaeological Science* 35, 947-955.

Affleck, T., Edwards, K.J., & Clarke, A. 1988: Archaeological and palynological studies at the Mesolithic pitchstone and flint site of Auchareoch, Isle of Arran. *Proceedings of the Society of Antiquaries of Scotland* 118, 37-59.

Alexander, D. 2000: Excavations of Neolithic pits, later prehistoric structures and a Roman temporary camp along the line of the A96 Kintore and Blackburn Bypass, Aberdeenshire. *Proceedings of the Society of Antiquaries of Scotland* 130, 11-75.

Allen, C., & Edwards, K.J. 1987: The Distribution of Lithic Materials of Possible Mesolithic Age on the Isle of Arran. *Glasgow Archaeological Journal* 14, 19-24.

Alsaker, S. 1987: *Bømlo - Steinalderens råstoffsentrum på Sørvestlandet*. Arkeologiske Avhandlinger 4. Bergen: Historisk Museum, Universitetet i Bergen.

Andersen, S.H. 1978: Flade, skælhuggede skiver af Brovsttype. Fremstillingsteknikken af de tidligste tværpile i Jylland. KUML 1978, 77-98.

Anderson, D.G. 2002: The Evolution of Tribal Social Organization in the Southeastern United States. *In* W.A. Parkinson (ed.): *The Archaeology of Tribal Societies*. Archaeological Series 15, 246-277. Ann Arbor: International Monographs in Prehistory.

Ashmore, P. 2005: Dating Barnhouse. *In* C. Richards (ed.) 2005: *Dwelling among the monuments. The Neolithic village of Barnhouse, Maeshowe passage grave and surrounding monuments at Stenness, Orkney*, 385-388. McDonald Institute Monographs. Cambridge: McDonald Institute for Archaeological Research.

Ashton, N., Dean, P., & McNabb, J. 1991: Flaked flakes: what, when and why? *Lithics* 12, 1-11.

Atkinson, J. 2002: Excavation of a Neolithic occupation site at Chapelfield, Cowie, Stirling. *Proceedings of the Society of Antiquaries of Scotland* 132, 139-192.

Atkinson, J.A., Dalglish, C., McLellan, K., & Swan, D. 2005: *Ben Lawers Historic Landscape Project: Excavations at Kiltyrie and Tombreck 2005*. Unpublished report commissioned by GUARD.

Ballin, TB. 1996a: *Klassifikationssystem for Stenartefakter*. Universitetets Oldsaksamling, Varia 36. Oslo: Universitetets Oldsaksamling.

Ballin, T.B. 1996b: Mikroliter. Diskussion af et begreb. *Universitetets Oldsaksamling, Årbok* 1995/1996, 7-14.

Ballin, T.B. 1999: *Kronologiske og Regionale Forhold i Sydnorsk Stenalder. En Analyse med Udgangspunkt i Bopladserne ved Lundevågen (Farsundprosjektet)*. Unpublished PhD thesis, Institute of Prehistoric Archaeology, Aarhus University.

Ballin, T.B. 2000: *The Lithic Assemblage from Rosinish, Benbecula, Western Isles*. Unpublished report commissioned by Historic Scotland.

Ballin, T.B. 2002a: Later Bronze Age Flint Technology: A presentation and discussion of post-barrow debitage from monuments in the Raunds area, Northamptonshire. *Lithics* 23, 3-28.

Ballin, T.B. 2002b: *The Lithic Assemblage from Achnahaird Sands, Highland*. Unpublished report commissioned by SUAT Ltd.

Ballin, T.B. 2002c: *The Lithic Assemblage from Dalmore, Isle of Lewis, Western Isles*. Unpublished report commissioned by Cardiff University.

Ballin, T.B. 2004a: The Mesolithic Period in Southern Norway: Material Culture and Chronology. *In* A. Saville (ed.): *Mesolithic Scotland and its Neighbours. The Early Holocene Prehistory of Scotland, its British and Irish Context, and some Northern European Perspectives*, 413-438. Edinburgh: Society of Antiquaries of Scotland.

Ballin, T.B. 2004b: The worked quartz vein at Cnoc Dubh, Isle of Lewis, Western Isles. Presentation and discussion of a small prehistoric quarry. *Scottish Archaeological Internet Reports (SAIR)* 11. [http://www.sair.org.uk/sair11/index.html.

Ballin, T.B. 2005a: *The Lithic Assemblage from East Lochside, Kirriemuir, Angus*. Unpublished report commissioned by CFA Archaeology Ltd.

Ballin, T.B. 2005b: *The Lithic Assemblage from Urquhart Castle Visitor Centre, Highland*. Unpublished report commissioned by GUARD.

Ballin, T.B. 2005c: Re-Examination of the Quartz Artefacts from Scord of Brouster. A lithic assemblage from Shetland and its Neolithic context. *Scottish Archaeological Internet Reports* 17. [http://www.sair.org.uk/sair17/index.html].

Ballin, T.B. 2006a: Re-examination of the Early Neolithic pitchstone-bearing assemblage from Auchategan, Argyll, Scotland. *Lithics* 27.

Ballin, T.B. 2006b: The Scottish Archaeological Pitchstone

Project (SAPP). *The Newsletter of the Society of Antiquaries of Scotland* 18 (2), 4-5.

Ballin, T.B. 2007a: The Scottish Archaeological Pitchstone Project. *IAOS (International Association for Obsidian Studies) Bulletin* 37 (Summer 2007), 13-15.

Ballin, T.B. 2007b: The Territorial Structure in the Stone Age of Southern Norway. *In* C. Waddington & K. Pedersen (eds.): *The Late Palaeolithic and Mesolithic of the North Sea Basin and Littoral*, Proceedings from a Conference at the University of Newcastle-upon-Tyne, 17 May 2003, 114-136. Oxford: Oxbow Books.

Ballin, T.B. 2008a: The distribution of Arran pitchstone – territories, exchange and the 'English Problem'. *PAST* 59.

Ballin, T.B. 2008b: The Lithic Assemblage. *In* G. Mudie: Excavations on the site of a late Iron Age roundhouse and souterrain, Glen Cloy, Brodick, Isle of Arran, 20-22. *Scottish Archaeological Journal* 29.1 (2007), 1-30.

Ballin, T.B. 2008c: *The Lithic Assemblage from Home Farm, Portree, Isle of Skye*. Unpublished report commissioned by CFA Archaeology Ltd.

Ballin, T.B. 2008d: *The Lithic Assemblage from Home Farm, Portree, Isle of Skye*. Unpublished report commissioned by Headland Archaeology Ltd.

Ballin, T.B. 2008e: Scottish Archaeological Pitchstone. *Archaeology Scotland*, 1, 6-7.

Ballin, T.B. forthcoming a: The British Late Neolithic 'Levalloisian', and other operational schemas from the later prehistoric period. A discussion based on finds from the Stoneyhill Project, Aberdeenshire. *Proceedings of Conference held by the British Neolithic Studies Group, at the British Museum 2005*.

Ballin forthcoming b: Detailed characterisation and discussion of the pitchstone artefacts from Barnhouse in the light of recent research into Scottish archaeological pitchstone. *The New Orcadian Antiquarian Journal*.

Ballin, T.B. forthcoming c: The Flint Assemblage. *In* J. Harding & F. Healy: *Raunds Area Project. The Neolithic and Bronze Age Landscapes of West Cotton, Stanwick and Irthlingborough, Northamptonshire*. English Heritage Archaeological Reports. London: English Heritage.

Ballin, T.B. forthcoming d: The Lithic Assemblage. *In* G. MacGregor (ed.): *Midross, Loch Lomond, Argyll*.

Ballin, T.B. forthcoming e: The Lithic Assemblage. *In* J.C. & H.K. Murray: Garthdee Road, Aberdeen City, Aberdeenshire. *Proceedings of the Society of Antiquaries of Scotland*.

Ballin, T.B. forthcoming f: The Lithic Assemblage. *In* E. Proudfoot & R. Turner: Fordhouse Barrow, House of Dun, Angus. *Proceedings of the Society of Antiquaries of Scotland*.

Ballin, T.B. forthcoming g: The Lithic Assemblage. *In* I. Suddaby: Stoneyhill Landfill Site, Peterhead, Aberdeenshire. *Proceedings of the Society of Antiquaries of Scotland*.

Ballin, T.B. forthcoming h: Lundevågen 31, Vest-Agder, SW Norway. The spatial organization of small hunter-gatherer sites – a case study (or: Binford in Practice). *In*

C. Bond: *Lithic Technology: Reduction and Replication*. Occasional Paper for the Lithic Studies Society. Oxbow Books.

Ballin, T.B. forthcoming i: Overhowden and Airhouse – Late Neolithic exotic flint, with focus on the period's exchange patterns. *Proceedings of the Society of Antiquaries of Scotland*.

Ballin, T.B. forthcoming j: Quartz Technology in Scottish Prehistory. *Scottish Archaeological Internet Reports (SAIR)*.

Ballin, T.B., & Faithfull, J. forthcoming: Gazetteer of Arran Pitchstone Sources. Presentation of exposed pitchstone dykes and sills across the Isle of Arran, and discussion of the possible archaeological relevance of these outcrops. *Scottish Archaeological Internet Reports (SAIR)*.

Ballin, T.B. & Johnson, M. 2005: A Mesolithic Chert Assemblage from Glentaggart, South Lanarkshire, Scotland: Chert Technology and Procurement Strategies. *Lithics* 26, 57-86.

Ballin, T.B., & Lass Jensen, O. 1995: *Farsundprosjektet. Stenalderbopladser på Lista*. Universitetets Oldsaksamling, Varia 29. Oslo: Universitetets Oldsaksamling.

Ballin, T.B., & Ward, T. 2008: General characterisation of the Biggar pitchstone artefacts, and discussion of Biggar's role in the distribution of pitchstone across Neolithic northern Britain. [http://www.biggararchaeology.org.uk/projects.php]

Ballin, T.B., Barrowman, C., & Faithfull, J. forthcoming: The unusual pitchstone-bearing assemblage from Blackpark Plantation East, Bute. *Transactions from the Buteshire Natural History Society*.

Banks, I. 1995: The excavation of three cairns at Stoneyburn Farm, Crawford, Lanarkshire, 1991. *Proceedings of the Society of Antiquaries of Scotland* 125, 289-343.

Barber, J. (ed.) 1997: *The Archaeological Investigation of a Prehistoric Landscape: Excavations on Arran 1978-1981*. STAR Monograph 2. Edinburgh: Scottish Trust for Archaeological Research.

Barclay, G.J. 1983: Sites of the third millennium BC to the first millennium AD at North Mains, Strathallan, Perthshire. *Proceedings of the Society of Antiquaries of Scotland* 113, 122-281.

Barclay, G.J., & Russell-White, C.J. 1993: Excavations in the ceremonial complex of the fourth to second millennium BC at Balfarg/Balbirnie, Glenrothes, Fife. *Proceedings of the Society of Antiquaries of Scotland* 123, 43-210.

Barclay, G.J., & Wickham-Jones, C.R. 2002: The investigation of some lithic scatters in Perthshire. *Tayside and Fife Archaeological Journal* 8, 1-9.

Barclay, G.J., Brophy, K., & MacGregor, G. 2002: Claish, Stirling: an early Neolithic structure in its context. *Proceedings of the Society of Antiquaries of Scotland* 132, 65-137.

Barclay, G.J., Carter, S.J., Dalland, M.M., Hastie, M., Holden, T.G., MacSween, A., & Wickham-Jones, C.R. 2001: A possible Neolithic settlement at Kinbeachie,

Black Isle, Highland. *Proceedings of the Society of Antiquaries of Scotland* 131, 57-85.

Beck, C., & Shennan, S. 1991: *Amber in Prehistoric Britain*. Oxbow Monograph 8. Oxford: Oxbow Books.

Becker, C.J. 1952: Die Nordschwedischen Flintdepots. *Acta Archaeologica* XXIII, 31-78.

Bender, B. 1985: Emergent Tribal Formation in the American Midcontinent. *American Antiquity* 50(1), 52-62.

Bergsvik, K.A., & Bruen Olsen, A. 2003: Traffic in Stone Adzes in Mesolithic Western Norway. *In* L. Larsson, H. Kindgren, K. Knutsson, D. Loeffler, & A. Åkerlund (eds.) 2003: *Mesolithic on the Move. Papers presented at the Sixth International Conference on the Mesolithic in Europe, Stockholm 2000*, 395-404. Oxford: Oxbow Books.

Binford, L.R. 1976: Forty-seven Trips: a case study in the character of some formation processes of the archaeological record. *In* Hall, E.S. (ed.): *Contributions to Anthropology: The Interior Peoples of Northern Alaska*, 299-351. National Museum of Man, Mercury Series 49. Ottawa.

Binford, L.R. 1978: Dimensional Analysis of Behaviour and Site Structure: Learning from an Eskimo Hunting Stand. *American Antiquity* 43 (3), 330-361.

Binford, L.R. 1983: *In Pursuit of the Past. Decoding the Archaeological Record*. London: Thames & Hudson.

Blundell, V., & Layton, R. 1978: Marriage, Myth and Models of Exchange in the West Kimberleys. *In* J. Specht & J.P. White (eds): *Trade and Exchange in Oceania and Australia*. Mankind 11(3), 231-245.

Bradley, R., & Edmonds, M. 1993: *Interpreting the Axe Trade: Production and Exchange in neolithic Britain*. Cambridge: Cambridge University Press.

Braudel, F. 1980: History and the Social Sciences: the *Longue Durée*. *In* F. Braudel & S. Matthews: *On History*, 25-54. Chicago: University of Chicago Press.

Braun, D., & Plog, S. 1982: Evolution of 'Tribal' Social Networks: Theory and Prehistoric North American Evidence. *American Antiquity* 47, 504-527.

Brinch Petersen, E. 1966: Klosterlund - Sønder Hadsund - Bøllund. Les trois sites principaux du Maglémosien ancien en Jutland, essai de typologie et de chronologie. *Acta Archaeologica* XXXVII, 77-185.

Brown, D.J., Bell, B.R., & Muirhead, D.K. 2007: *Silicic pyroclastic volcanism in the North Atlantic Igneous Province (NAIP): a re-interpretation of the Sgurr of Eigg Pitchstone, NW Scotland. Abstracts of the Volcanic Magmatic Studies Group, Winter Meeting, January 2007*. [http://www.davidjbrown.org.uk/presentations/eigg%20 vmsg%20oxford.pdf].

Bruen Olsen, A. 1992: *Kotedalen - en boplass gjennom 5000 år. Bind 1. Fangstbosetning og tidlig jordbruk i vestnorsk steinalder: Nye funn og nye perspektiver*. Bergen: Historisk Museum, Universitetet i Bergen.

Bruen Olsen, A., & Alsaker, S. 1984: Greenstone and Diabase Utilization in the Stone Age of Western Norway: Technological and Socio-cultural Aspects of Axe and Adze Production and Distribution. *Norwegian Archaeological Review* 17 (2), 71-103.

Bryce, J. 1859: *Geology of Clydesdale and Arran*. London, Glasgow: Richard Griffin and Company.

Bryce, J. 1862: An Account of Excavations within the Stone Circles of Arran. *Proceedings of the Society of Antiquaries of Scotland* 4 (1860-62), 499-524.

Bryce, T.H. 1902: On the Cairns of Arran – A Record of Explorations with an Anatomical Description of the Human Remains Discovered. *Proceedings of the Society of Antiquaries of Scotland* 36 (1901-02), 74-181.

Bryce, T.H. 1903: On the Cairns of Arran - A Record of Further Explorations during the Season of 1902. *Proceedings of the Society of Antiquaries of Scotland* 37 (1902-03), 36-67.

Bryce, T.H. 1904: On the cairns and tumuli of the Island of Bute. A record of explorations during the season of 1903. *Proceedings of the Society of Antiquaries of Scotland* 38 (1903-4), 17-81.

Bryce, T.H. 1909: On the Cairns of Arran. No. III. With a Notice of a Megalithic Structure at Ardnadam, on the Holy Loch. *Proceedings of the Society of Antiquaries of Scotland* 43 (1908-09), 337-370.

Bryce, T.H. 1910: The Sepulchral Remains. *In* J.A. Balfour (ed.)*: The Book of Arran*, 33-155. Glasgow: Archaeological Society of Glasgow.

Burgess, C.B. 1972: Goatscrag: a Bronze Age rock shelter cemetery in North Northumberland. *Archaeologia Aeliana* 4, Ser 50, 15-69.

Callander, J.G. 1928: A Collection of Stone Implements from Airhouse, Parish of Channelkirk, Berwickshire. *Proceedings of the Society of Antiquaries of Scotland* LXII (1927-28), 166-180.

Caldwell, J.R. 1964: Interaction Spheres in Prehistory. *In* J.R. Caldwell & R.L. Hall (eds): *Hopewellian Studies*. Illinois State Museum, Scientific Papers 12(6), 133-143. Springfield: Illinois State Museum.

Carneiro, R.L. 2002: The Tribal Village and Its Culture: An Evolutionary Stage in the History of Human Society. *In* W.A. Parkinson (ed.): *The Archaeology of Tribal Societies*. Archaeological Series 15, 34-52. Ann Arbor: International Monographs in Prehistory.

Clark, J.G.D. 1934a: The Classification of a Microlithic Culture: The Tardenoisian of Horsham. *The Archaeological Journal* XC, 52-77.

Clark, J.G.D. 1934b: Derivative Forms of the Petit Tranchet in Britain. *The Archaeological Journal* XCI, 32-58.

Clark, J.G.D. 1975: *The Earlier Stone Age Settlement of Scandinavia*. Cambridge: Cambridge University Press.

Clarke, D.L. 1968: *Analytical Archaeology*. London: Methuen.

Clarke, A. 1989: Corse Law, Carnwath: a lithic scatter. *Proceedings of the Society of Antiquarians of Scotland* 119, 43-54.

Clarke, D.V., & Sharples, N. 1985: Settlement and Subsistence in the Third Millennium BC. *In* C. Renfrew (ed.): *The Prehistory of Orkney*, 54-82. Edinburgh: Edinburgh University Press.

Clough, T.H.M., & Cummins, W.A. (eds.) 1979: *Stone Axe Studies. Archaeological, Petrological, Experimental and Ethnographic*. CBA Research Reports 23. London: Council for British Archaeology.

Clough, T.H.M., & Cummins, W.A. (eds.) 1988: *Stone Axe Studies, 2. The Petrology of Prehistoric Stone Implements from the British Isles*. CBA Research Reports 67. London: Council for British Archaeology.

Coles, J.M., & Simpson, D.D.A. 1965: The Excavation of a Neolithic Round Barrow at Pitnacree, Perthshire, Scotland. *Proceedings of the Prehistoric Society* XXXI, 34-57.

Conway 1994: *'Shandragh', Knockans South, Rathlin Island. Late Neolithic settlement and industrial site*. Antrim 1994:008, D130515, SMR 1:82. [http://www.excavations.ie]

Cooney, G., & Mandal, S. 1998: *The Irish Stone Axe Project, Monograph I*. Bray: Wordwell Ltd.

Corcoran, J.X.W.P. 1966: Excavation of three chambered cairns at Loch Calder, Caithness. *Proceedings of the Society of Antiquaries of Scotland* 98 (1964-66), 1-75.

Cormack, W.F. 1964: Burial site at Kirkburn, Lockerbie. *Proceedings of the Society of Antiquaries of Scotland* 96 (1963-4), 107-135.

Cormack, W.F. 1970: A Mesolithic Site at Barsalloch, Wigtownshire. *Transactions of the Dumfriesshire & Galloway Natural History & Antiquarian Society* 47, 63-80.

Cowie, T.G. 1996: Torrs Warren, Luce Sands, Galloway: a report on archaeological and palaeoecological investigations undertaken in 1977 and 1979. *Transactions of the Dumfriesshire & Galloway Natural History & Antiquarian Society* LXXI, 11-105.

Cummings, V. 2002: Between Mountains and Sea: a Reconsideration of the Neolithic Monumenets of South-West Scotland. *Proceedings of the Prehistoric Society* 68, 125-146.

Cummings, V., & Fowler, C. 2007: *From Cairn to Cemetery. An archaeological investigation of the chambered cairns and early Bronze Age mortuary deposits at Cairnderry and Bargrennan White Cairn, south-west Scotland*. BAR British Series 434. Oxford: BAR Publishing.

Cummins, W.A. 1979: Neolithic stone axes: distribution and trade in England and Wales. *In* T.H.M. Clough & W.A. Cummins (eds.) 1979: *Stone Axe Studies. Archaeological, Petrological, Experimental and Ethnographic*, 5-12. CBA Research Reports 23. London: Council for British Archaeology.

Darvill, T. 1995: *Billown Neolithic Landscape Project, Isle of Man, 1995*. Bournemouth University School of Conservation Sciences, Research Report 1. Bournemouth / Douglas: School of Conservation Sciences, Bournemouth University / Manx National Heritage.

Dalton, G., 1977: Aboriginal Economies in Stateless Societies. *In* T.K. Earle & J.E. Ericson (eds.) 1977: *Exchange Systems in Prehistory*, 191-212. New York: Academic Press.

Davidson, D.A., & Carter, S.P. 1997: Soils and Their Evolution. *In* K.J. Edwards & I.B.M. Ralston, I.B.M. 1997:

Scotland: Environment and Archaeology, 8000 BC - AD 1000, 45-62. Chicester: John Wiley & Sons.

Donnelly, M. 2002: Struck stone. *In* J. Atkinson: Excavation of a Neolithic occupation site at Chapelfield, Cowie, Stirling. *Proceedings of the Society of Antiquaries of Scotland* 132, 169-173.

Donnelly, M., & Finlay, N. forthcoming: *The Arran Ring Main Water Pipeline: Lithic Report*.

Durden, T. 1995: The production of specialised flintwork in the later Neolithic: a case study from the Yorkshire Wolds. *Proceedings of the Prehistoric Society* 61, 409-432.

Earle, T.K., & Ericson, J.E. (eds) 1977a: *Exchange Systems in Prehistory*. New York: Academic Press.

Earle, T.K., & Ericson, J.E. 1977b: Exchange Systems in Archaeological Perspective. *In* T.K. Earle & J.E. Ericson (eds.): *Exchange Systems in Prehistory*, 3-12. New York: Academic Press.

Emeleus, C.H., & Bell, B.R. 2005: *The Palaeogene volcanic districts of Scotland*. British Regional Geology. Nottingham: British Geological Survey.

Ericson, J.E. 1982: Production for Obsidian Exchange in California. *In* J.E. Ericson & T.K. Earle (eds): *Contexts for Prehistoric Exchange*, 129-148. New York: Academic Press.

Evans-Pritchard, E.E. 1940: The Nuer. Oxford, New York: Oxford University Press.

Fell, C.I. 1990: Prehistoric Flint and Stone Material. *In* M.R. McCarthy: *A Roman, Anglian and Medieval Site at Blackfriars Street, Carlisle: Excavations 1977-79*, 91-97. Cumberland & Westmorland Antiquarian & Archaeological Society Research Series 4. Kendal: Cumberland & Westmorland Antiquarian & Archaeological Society.

Finlay, N. 1997: Various entries in: J. Barber (ed.): *The Archaeological Investigation of a Prehistoric Landscape: Excavations on Arran 1978-1981*. STAR Monograph 2. Edinburgh: Scottish Trust for Archaeological Research.

Fischer, A., Grønnow, B., Jønsson, J.H., Nielsen, F.O., & Petersen, C. 1979: *Stenaldereksperimenter i Lejre. Bopladsernes indretning*. Working Papers, The National Museum of Denmark 8. København: The National Museum of Denmark.

Fisher, L.E., & Eriksen, B.V. (eds.) 2002: *Lithic Raw Material Economies in Late Glacial and Early Postglacial Europe*. British Archaeological Reports International Series 1093. Oxford: British Archaeological Reports.

Ford, R.I. 1972: Barter, Gift, or Violence: An Analysis of Tewa Intertribal Exchange. *In* E.N. Wilmsen (ed.): *Social Exchange and Interaction*, 21-46. Museum of Anthropology, University of Michigan, Anthropological Papers 46. Ann Arbor: University of Michigan.

Foster, S.M. 2006: *Maeshowe and the Heart of Neolithic Orkney*. Edinburgh: Historic Scotland.

Fowles, S.M. 2002: From Social Type to Social Process: Placing 'Tribe' in a Historical Framework. *In* W.A. Parkinson (ed.): *The Archaeology of Tribal Societies*. Archaeological Series 15, 13-33. Ann Arbor: International Monographs in Prehistory.

Fraser, S., & Murray, H. 2005: New Light on the Earliest Neolithic in the Dee Valley, Aberdeenshire. *PAST* 50, 1-2.

Gebauer, A.B. 1987: Stylistic Analysis. A Critical Review of Concepts, Models, and Interpretations. *Journal of Danish Archaeology* 6, 223-229.

Gendel, P.A. 1984: *Mesolithic Social Territories in Northwestern Europe*. BAR, International Series 218. Oxford: BAR.

Geneste, J.-M. 1988a: Les industries de la Grotte Vaufrey: technologie du débitage, économie et circulation de la matière première lithique. *In* J.-P. Rigaud (ed.): *La Grotte Vaufrey à Cenac et Saint-Julien (Dordogne), Paléoenvironments, Chronologie et Activités humaines.* Mémoires de la Société Préhistorique Française 19, 441-518.

Geneste, J.-M. 1988b: Systèmes d'approvisionnement en matières premières au paléolithique moyen et au paléolithique supérieur en Aquitaine. *L'Homme de Néanderthal* 8, 61-70.

Goldstein, L.G. 1976: *Spatial Structure and Social Organization: Regional Manifestations of Mississippian Society*. Ann Arbor: Unpublished Ph.D. dissertation from University of Michigan.

Gould, R.A. 1980: *Living Archaeology*. Cambridge: Cambridge University press.

Green, H.S. 1980: *The Flint Arrowheads of the British Isles. A detailed study of material from England and Wales with comparanda from Scotland and Ireland*. BAR British Series 75(i). Oxford: BAR.

Gunn, W., Geikie, A., Peach, B.N., & Harker, A. 1903: *Scotland. The Geology of North Arran, South Bute, and the Cumbraes, with Parts of Ayrshire and Kintyre*. Memoirs of the Geological Survey. Glasgow: James Hedderwick & Sons / His Majesty's Stationery Office.

Haggarty, A. 1991: Machrie Moor, Arran: recent excavations at two stone circles. *Proceedings of the Society of Antiquaries of Scotland* 121, 51-94.

Hawkes, C. 1954: Archaeological Theory and Method: Some Suggestions from the Old World. *American Anthropologist* 56, 155-168.

Healy, F. 1993: Lithic Material. *In*: R. Bradley, P. Chowne, R.M.J. Cleal, F. Healy, & I. Kinnes: *Excavations on Redgate Hill, Hunstanton, Norfolk, and at Tattershall Thorpe, Lincolnshire*. East Anglian Archaeology Report 57, 28-39. Gressenhall: Field Archaeology Division, Norfolk Museums Service / Heritage Trust of Lincolnshire.

Helm, J. 1973: The Nature of Dogrib Socioterritorial Groups. *In* R.B. Lee og I. DeVore (eds.): *Man the Hunter*, 118-125. Chicago: Aldine

Henshall, A.S. 1963: *The Chambered Tombs of Scotland. Volume One*. Edinburgh: Edinburgh University Press.

Henshall, A.S. 1972: *The Chambered Tombs of Scotland. Volume Two*. Edinburgh: Edinburgh University Press.

Herity, M., & Eogan, G. 1996: *Ireland in Prehistory*. London: Routledge.

Higgs, E.S., & Vita-Finzi, C. 1972: Prehistoric economies: a territorial approach. *In* E.S. Higgs (ed.): *Papers in economic prehistory*, 27-36. Cambridge: Cambridge University Press.

Hodder, I. 1974: Regression analysis of some trade marketing patterns. *World Archaeology* 6 (1), 172-189.

Hodder, I. 1979: Economic and Social Stress and Material Culture Patterning. *American Antiquity* 44 (3), 446-454.

Houlder, C.H. 1979: The Langdale and Scafell Pike axe factory sites: a field survey. *In* T.H.M. Clough & W.A. Cummins (eds.) 1979: *Stone Axe Studies. Archaeological, Petrological, Experimental and Ethnographic*, 87-89. CBA Research Reports 23. London: Council for British Archaeology.

Högberg, A., & Olausson, D. 2007: *Scandinavian Flint - an Archaeological Perspective*. Aarhus: Aarhus Universitetsforlag.

Inizan, M.-L., Roche, H., & Tixier, J. 1992: *Technology of Knapped Stone*. Préhistoire de la Pierre Taillée 3. Meudon: Cercle de Recherches et d'Etudes Préhistoriques.

Jameson, R. 1798: An Outline of the Mineralogy of the Shetland Islands and the Island of Arran. Edinburgh: W. Creech.

Jarman, M.R. 1972: A Territorial Model for Archaeology: A Behavioral and Geographical Approach. *In* D. Clarke (ed.): *Models in Archaeology*, 705-733. London: Methuen.

Jensen, J. 2001: *Danmarks Oldtid. Stenalder 13.000 - 2.000 f. Kr.* København: Gyldendal.

Johnson, M., & Ballin, T.B. 2006: Gaining Knowledge from the Ploughsoil: A Finds scatter from East Lochside, Kirriemuir. *Scottish Archaeology News* 51, 9.

Johnston, D. 1997: Biggar Common 1987-93: an early prehistoric funerary and domestic landscape in Clydesdale, South Lanarkshire. *Proceedings of the Society of Antiquaries of Scotland* 127, 185-253.

Jones, A. 1997: On the earth-colours of Neolithic death. *British Archaeology*, 22. [http://www.britarch.ac.uk/BA/ba22/ba22feat.html]

Judd, J.W. 1893: On Composite Dykes in Arran. *Quarterly Journal of the Geological Society of London* XLIX, 536-564.

Juel Jensen, H. 1994: *Flint Tools and Plant Working. Hidden Traces of Stone Age Technology. A use wear study of some Danish Mesolithic and TRB implements*. Århus: Aarhus University Press.

Kindgren, H. 1991: Kambrisk flinta och etniska grupper i Västergötlands senmesolitikum. *In* H. Browall, P. Persson, & K.G. Sjögren (eds.): *Västsvenska stenåldersstudier, vol. 1*. Gotarc, serie c, Arkeologiska skrifter 8, 71-110. Göteborg: Institutionen för arkeologi, Göteborgs universitet.

Kinnes, I. 1985: Circumstance not context: the Neolithic of Scotland as seen from outside. *Proceedings of the Society of Antiquaries of Scotland* 115, 15-57.

Lacaille, A.D. 1930: Mesolithic Implements from Ayrshire. *Proceedings of the Society of Antiquaries of Scotland* LXIV, 34-48.

Lacaille, A.D. 1931: A Bronze Age Cemetery near Cowdenbeath, Fife. *Proceedings of the Society of Antiquaries of Scotland* 65 (1930-31), 261-269.

Lacaille, A.D. 1937: The Microlithic Industries of Scotland. *Transactions of the Glasgow Archaeological Society* IX (I), 56-74.

Lacaille, A.D. 1945: The Stone Industries Associated with the Raised Beach at Ballantrae. *Proceedings of the Society of Antiquaries of Scotland* 79 (1944-45), 81-106.

Larsson, L. 2004: Axeheads and fire – the transformation of wealth. *In* E.A. Walker, F. Wenban-Smith, & F. Healy (eds.) 2004: *Lithics in Action. Papers from the Conference Lithic Studies in the Year 2000.* Oxbow Books / Lithic Studies Society Occasional Paper 8. Oxford: Oxbow Books / Lithic Studies Society.

Lebour, N. 1914: White Quartz Pebbles and their Archaeological Significance. *Transactions of the Dumfriesshire & Galloway Natural History & Antiquarian Society* 1913-14, 121-134.

Lelong, O., & Pollard, T. 1998: Excavations of a Bronze Age ring cairn at Cloburn Quarry, Cairngryff Hill, Lanarkshire. *Proceedings of the Society of Antiquaries of Scotland* 128 105-142.

Le Maitre, R.W. (ed.) 2002: *Igneous Rocks: A Classification and Glossary of Terms. Recommendations of the International Union of Geological Sciences Subcommission on the Systematics of Igneous Rocks.* 2nd Edition. Cambridge: Cambridge University Press.

Lemonnier, P. 1976. La Description des Chaînes Opératoires: Contribution à l'Analyse des Systèmes Techniques. *Technique et Culture* 1. 100-151.

Leroi-Gourhan, A. 1965. *Le Geste et la Parole II. La Memoire et les Rythmes.* Paris: Albin Michel.

MacKie, E. 1964: New Excavations on the Monamore Neolithic Chambered Cairn, Lamlash, Isle of Arran, in 1961. *Proceedings of the Society of Antiquaries of Scotland* 97 (1963-64), 1-34.

MacKie, E.W. 1973: Duntreath. *Current Archaeology* 36, 6-7.

Manby, T.G. 1979: Typology, materials, and distribution of flint and stone axes in Yorkshire. *In* T.H.McK Clough & W.A. Cummins (eds.): *Stone Axe Studies. Archaeological, Petrological, Experimental and Ethnographic.* CBA Research Report 23. London: Council for British Archaeology, 65-81.

Mann, L.M. 1918: The Prehistoric and Early Use of Pitchstone and Obsidian. *Proceedings of the Society of Antiquaries of Scotland* LII, 140-149.

Marshall, D.N. 1978: Excavations at Auchategan, Glendaruel, Argyll. *Proceedings of the Society of Antiquaries of Scotland* 109, 36-74.

Masters, L. 1997: The excavation and restoration of Camster Long chambered cairn, Caithness, Highland, 1967-80. *Proceedings of the Society of Antiquaries of Scotland* 127, 123-183.

Maynard, D. 1993: Neolithic Pit at Carzield, Kirkton, Dumfriesshire. *Transactions of the Dumfriesshire & Galloway Natural History & Antiquarian Society* LXVIII, 25-32.

McCartan, S.B., & Johnson, A. 1991: A Rescue Excavation at Ballachrink, Jurby. *Proceedings of the Isle of Man Natural History and Antiquarian Society* X (1), 105-122.

Mercer, J. 1968: Stone Tools from a Washing-Limit Deposit of the Highest Post-Glacial Transgression, Lealt Bay, Isle of Jura. *Proceedings of the Society of Antiquaries of Scotland* 100 (1967-68), 1-46.

Mercer, J. 1971: A Regression-time Stone-workers' Camp, 33 ft OD, Lussa River, Isle of Jura. *Proceedings of the Society of Antiquaries of Scotland* 103, 1-32.

Mercer, J. 1972: Microlithic and Bronze Age Camps, 75-26 ft OD, N Carn, Isle of Jura. *Proceedings of the Society of Antiquaries of Scotland* 104, 1-22.

Mercer, J. 1980: Lussa Wood 1: The Late Glacial and Early Post-Glacial Occupation of Jura. *Proceedings of the Society of Antiquaries of Scotland* 110, 1-32.

Middleton, R. 2005: The Barnhouse Lithic Assemblage. *In* C. Richards (ed.) 2005: *Dwelling among the monuments. The Neolithic village of Barnhouse, Maeshowe passage grave and surrounding monuments at Stenness, Orkney*, 293-321. McDonald Institute Monographs. Cambridge: McDonald Institute for Archaeological Research.

Mitchell, A. 1898: James Robertson's Tour through some of the Western Isles &c., of Scotland in 1768. *Proceedings of the Society of Antiquaries of Scotland* 32 (1897-98), 11-19.

Morgan, L.H. 1995: *League of the Iroguois.* North Dighton, MA: JG Press. Reprint of the original from 1851.

Morrow, C.A., & Jefferies, R.W. 1989: Trade or embedded procurement? A test case from southern Illinois. *In* R. Torrence (ed.): *Time, Energy and Stone Tools*, 27-33. Cambridge: Cambridge University Press.

Mulholland, H. 1970: The Microlithic Industries of the Tweed Valley. *Transactions of the Dumfriesshire & Galloway Natural History & Antiquarian Society* 47, 81-110.

Murray, H., & Murray, C. forthcoming: Warrenfield, Crathes, Aberdeenshire.

Ness, J., & Ward, T. 2001: *Pitchstone Seminar held Saturday 30 September 2000. Report.* Biggar: Biggar Museum Trust.

Orton, C. 1980: *Mathematics in Archaeology.* London: Collins.

O'Shea, J.M., & Milner, C.M. 2002: Material Indicators of Territory, Identity, and Interaction in a Prehistoric Tribal System. *In* W.A. Parkinson (ed.): *The Archaeology of Tribal Societies.* Archaeological Series 15, 200-226. Ann Arbor: International Monographs in Prehistory.

Orme, B. 1981: *Anthropology for Archaeologists: An Introduction.* London: Duckworth.

Parkinson, W.A. 2002: Introduction: Archaeology and Tribal Societies. *In* W.A. Parkinson (ed.): *The Archaeology of Tribal Societies.* Archaeological Series 15, 1-12. Ann Arbor: International Monographs in Prehistory.

Peltenburg, E.J. 1982: Excavations at Balloch Hill, Argyll. *Proceedings of the Society of Antiquaries of Scotland* 112, 142-214.

Petrequin, P., Cassen, S., Croutsch, C., & Errera, M. 2002: La valorisation sociale des longues haches dans l'Europe néolithique. *In* J. Guilaine (ed.): *Matériaux, productions, circulations de Néolitihique à l'Age du Bronze*, 67-98. Paris: Editions Errance.

Piggott, S., & Powell, T.G.E. 1951: The excavation of three Neolithic chambered cairns in Galloway, 1949. *Proceedings of the Society of Antiquaries of Scotland* 83 (1948-9), 123-129

Plog, F. 1977: Modeling Economic Exchange. *In* T.K. Earle, & J.E. Ericson (eds.): *Exchange Systems in Prehistory*. Studies in Archaeology, 127-140. New York: Academic Press.

Preston, R.J., Hole, M.J., Still, J., & Patton, H. 1998: The mineral chemistry and petrology of Tertiary pitchstones from Scotland. *Transactions of the Royal Society of Edinburgh: Earth Sciences* 89, 95-11.

Preston, J., Meighan, I., Simpson, D., & Hole, M. 2002: Mineral Chemical Provenance of Neolithic Pitchstone Artefacts from Ballygalley, County Antrim, Northern Ireland. *Geoarchaeology* 17 (3), 219-236.

Pryor, F. 1978: *Excavations at Fengate, Peterborough, England: The Second Report*. Royal Ontario Museum Archaeology Report 5. Toronto: Royal Ontario Museum.

Radcliffe-Brown, A.R. 1948: *The Andaman Islanders*. Glencoe, IL: The Free Press.

Renfrew, C. 1976: Megaliths, territories and populations. *In* S.J. de Laet (ed.): *Acculturation and Continuity in Atlantic Europe*. Dissertationes Archaeologicae Gandenses XVI, 198-220. Bruge: de Tempel.

Renfrew, C. 1977: Alternative Models for Exchange and Spatial Distribution. *In* T.K. Earle, & J.E. Ericson (eds.): *Exchange Systems in Prehistory*. Studies in Archaeology, 71-90. New York: Academic Press.

Renfrew, C. 1993: Trade Beyond the Material. *In* C. Scarre, & F. Healy (eds.): *Trade and Exchange in Prehistoric Europe. Proceedings of a Conference held at the University of Bristol, April 1992*. Oxbow Monographs 33, 5-16. Oxford: Oxbow Books.

Renfrew, C., & Bahn, P. 1996: *Archaeology. Theory, Methods and Practice*. London: Thames & Hudson.

Renfrew, C., Dixon, J.E., & Cann, J.R. 1966: Obsidian and Early Cultural Contact in the Near East. *Proceedings of the Prehistoric Society* XXXII, 30-72.

Renfrew, C., Dixon, J.E., & Cann, J.R. 1968: Further Analysis of Near Eastern Obsidians. *Proceedings of the Prehistoric Society* XXXIV, 319-331.

Rennie, E.B. 1984: Excavations at Ardnadam, Cowal, Argyll, 1964-1982. *Glasgow Archaeological Journal* 11, 13-40.

Richards, C. (ed.) 2005: *Dwelling among the monuments. The Neolithic village of Barnhouse, Maeshowe passage grave and surrounding monuments at Stenness, Orkney*. McDonald Institute Monographs. Cambridge: McDonald Institute for Archaeological Research.

Richey, J.E. 1961: *Scotland: The Tertiary Volcanic Districts*. British Regional Geology 3. Edinburgh: Natural Environment Research Council. Institute of Geological Sciences. Geological Survey and Museum / Her Majesty's Stationery Office.

Rideout, J.S. 1997: Excavations of Neolithic enclosures at Cowie Road, Bannockburn, Stirling, 1984-5. *Proceedings of the Society of Antiquaries of Scotland* 127, 29-68.

Ritchie, P.R. 1968: The Stone Implement Trade in Third-millenium Scotland. *In* J.M. Coles, & D.D.A. Simpson (eds.): *Studies in Ancient Europe. Essays presented to Stuart Piggott*, 119-136. Leicester: Leicester University Press.

Ritchie, J.N.G. 1970: Excavation of the Chambered Cairn at Achnacreebeag. *Proceedings of the Society of Antiquaries of Scotland* 102 (1969-70), 31-55.

Ritchie, P.R., & Scott, J.G. 1988: The petrological identification of stone axes from Scotland. *In* T.H.M. Clough & W.A. Cummins (eds.): *Stone Axe Studies, 2. The Petrology of Prehistoric Stone Implements from the British Isles*, 85-91. CBA Research Reports 67. London: Council for British Archaeology.

Roe, D.E. 1981: *The Lower and Middle Palaeolithic Periods in Britain*. The Archaeology of Britain. London, Boston and Henley: Routledge & Kegan Paul.

Roy, A.E., McGrail, N., & Carmichael, R. 1963: A New Survey of the Tormore Circles. *Transactions of Glasgow Archaeological Society*, 15 (2), 59-67.

Sahlins, M. 1961: The Segmentary Lineage: An Organization of Predatory Expansion. *American Anthropologist* 63, 332-345.

Sahlins, M. 1972: *Stone Age Economics*. Chicago: Aldine Publishing.

Saville, A. 1981: *Grimes Graves, Norfolk. Excavations 1971/72: Volume II. The Flint Assemblage*. Department of the Environment Archaeological Reports 11. London: Her Majesty's Stationery Office.

Saville, A. 1999: A Cache of Flint Axeheads and Other Flint Artefacts from Auchenhoan, near Campbelltown, Kintyre, Scotland. *Proceedings of the Prehistoric Society* 65, 83-114.

Saville, A. 2002: Lithic Artefacts from Neolithic Causewayed Enclosures: Character and Meaning. *In* G. Varndell, & P. Topping (eds.): *Enclosures in Neolithic Europe. Essays on Causewayed and Non-Causewayed Sites*, 91-105. Oxford: Oxbow Books.

Saville, A. 2006: The Early Neolithic Lithic Assemblage in Britain: some Chronological Considerations. *In* P. Allard, F. Bostyn, & A. Zimmermann (eds.): *Contribution of Lithics to Early and Middle Neolithic Chronology in France and Neighbouring Regions*. BAR International Series 1494, 1-14. Oxford: British Archaeological Reports.

Saville, A., Ballin, T.B., & Ward, T. 2008: Howburn, near Biggar, South Lanarkshire: preliminary notice of a Scottish inland early Holocene lithic assemblage. *Lithics* 28, 41-49.

Scott, J.G. 1956: The Excavation of the Chambered Tomb at Brackley, Kintyre, Argyll. *Proceedings of the Society of Antiquaries of Scotland* 89 (1955-56), 22-54.

Scott, J.G. 1964: The chambered cairn at Beacharra, Kintyre, Argyll. *Proceedings of the Prehistoric Society* XXX, 134-58.

Scott, J.G. 1977: The Rothesay style of Neolithic pottery in Scotland. *In* D.N. Marshall & I.D. Taylor: The excavation of the chambered cairn at Glenvoidean, Isle of Bute. *Proceedings of the Society of Antiquaries of Scotland* 108 (1976-7), 26-37.

Scott, J.G. 1998: The Bronze Age Burials at Monybachach, Skipness. *The Kintyre Magazine* 24. [http://www.kintyremag.co.uk/1998/24/page3.html]

Seeman, M.F. 1995: When Words Are Not Enough. Hopewell Interregionalism and the Use of Material Symbols at the GE Mound. *In* M.S. Nassaney, & K.E. Sassaman (eds.): *Native American Interactions. Multiscalar Analyses and Interpretations in the Eastern Woodlands,* 122-143. Knoxville: University of Tennessee Press.

Service, E.R. 1971: *Primitive Social Organization. An Evolutionary Perspective.* Second Edition. New York: Random House.

Sharp, J. 1912: Notice of a Collection of Flint Arrowheads and Implements found on the Farm of Overhowden, in the Parish of Channelkirk, Berwickshire. *Proceedings of the Society of Antiquaries of Scotland* 46 (1911-12), 370-372.

Sharp, L. 1952: Steel Axes for Stone-Age Australians. *Human Organization* 11, 17-22.

Sharples, N.M. 1981: The excavation of a chambered cairn, the Ord North, at Lairg, Sutherland by J.X.W.P. Corcoran. *Proceedings of the Society of Antiquaries of Scotland* 111, 21-62.

Shepherd, W. 1972: *Flint: its origin, properties and uses.* London.

Sheridan, A. 1986: Porcellanite artifacts: a new survey. *Ulster Journal of Archaeology* 49, 19-32.

Sheridan, A. 2007: From Picardie to Pickering and Pencraig Hill? New information on the 'Carinated Bowl Neolithic' in northern Britain. *Proceedings of the British Academy* 144, 441-492.

Simpson, D., & Meighan, I. 1999: Pitchstone - a new trading material in Neolithic Ireland. *Archaeology Ireland* 13, 26-30.

Skinner, C.E. 1983: *Obsidian Studies in Oregon: An Introduction to Obsidian and An Investigation of Selected Methods of Obsidian Characterization Utilizing Obsidian Collected at Prehistoric Quarry Sites in Oregon.* Unpublished Master's Terminal Project. Eugene: University of Oregon.

Skinner, C.E. 1997: *Geoarchaeological and Geochemical Investigations of the Devil Point Obsidian Source, Willamette National Forest, Western Cascades, Oregon.* Report prepared for the Willamette National Forest, Eugene, Oregon. Corvallis: Northwest Research Obsidian Studies Laboratory, Oregon.

Smith, J. 1897: *Prehistoric Man in Ayrshire.* London: Elliot Stock.

Smith, I.F. 1979: The chronology of British stone implements. *In* T.H. McK Clough & W.A. Cummins (eds.): *Stone Axe Studies. Archaeological, Petrological, Experimental and Ethnographic,* 13-22. CBA Research Report 23. London: Council for British Archaeology.

Squair, R., & Jones, A. 2002: Prehistoric pottery. *In* J. Atkinson: Excavation of a Neolithic occupation site at Chapelfield, Cowie, Stirling. *Proceedings of the Society of Antiquaries of Scotland* 132, 155-162.

Srivastava, V. 2008: Political anthropology. [http://www.waterandfood.org/gga/Lecture%20Material/VSrivastava_PoliticalAnthropology.pdf]

Stables, D. 1996: South Mound Cairn, Houston. *In* D. Alexander (ed.): *Prehistoric Renfrewshire. Papers in Honour of Frank Newall,* 23-28. Edinburgh: Renfrewshire Local History Forum, Archaeology Section.

Steffen, A. 2005: *The Dome Fire Obsidian Study: Investigating the Interaction of Heat, Hydration, and Glass Geochemistry.* Dissertation Submitted in Partial Fulfillment of the Requirements for the Degree of Doctor of Philosophy, Anthropology. Albuquerque: The University of New Mexico.

Stevenson, R.B.K. 1948: 'Lop-sided' Arrow-heads. *Proceedings of the Society of Antiquaries of Scotland* LXXX (1946-48), 179-182.

Steward, J. 1955: *Theory of Culture Change.* Urbana: University of Illinois Press.

Stewart, R.M. 1994: Late Archaic through Late Woodland Excange in the Middle Atlantic Region. *In* T.G. Baugh & J.E. Ericson (eds): *Prehistoric Exchange Systems in North America,* 73-98. New York: Plenum Press.

Struever, S. 1972: The Hopewell Interaction Sphere in Riverine-Western Great Lakes Culture History. *In* M.P. Leone (ed.): *Contemporary Archaeology. A Guide to Theory and Contributions,* 303-315. Carbondale: Southern Illinois University Press.

Struever, S., & Houart, G.L. 1964: The Hopewell Interaction Sphere. *In* J.R. Caldwell & R.L. Hall (eds): *Hopewellian Studies.* Illinois State Museum, Scientific Papers 12(3), 85-106. Springfield: Illinois State Museum.

Struever, S., & Houart, G.L. 1972: An Analysis of the Hopewell Interaction Sphere. *In* E.N. Wilmsen (ed.): *Social Exchange and Interaction,* 47-80. Museum of Anthropology, University of Michigan, Anthropological Papers 46. Ann Arbor: University of Michigan.

Telford, D. 2002: The Mesolithic Inheritance: Contrasting Neolithic Monumentality in Eastern and Western Scotland. *Proceedings of the Prehistoric Society* 68, 289-315.

Thatcher, J.J. 2001: *The Distribution of Geologic and Artifact Obsidian from the Silver Lake / Sycan Marsh Geochemical Source Group, South-Central Oregon.* Unpublished thesis submitted to Oregon State University in partial fulfillment of the requirements for the degree of Master of Arts in Interdisciplinary Studies. Corvallis: Oregon State University.

Tolan-Smith, C. 2001: *The Caves of Mid Argyll. An archaeology of human use.* Society of Antiquaries of Scotland Monograph Series 20. Edinburgh: Society of Antiquaries of Scotland.

Topping, P. 2005: Shaft 27 Revisited: An Ethnography. *In* Topping, P., & Lynott, M. (eds.) 2005: *The Cultural Landscape of Prehistoric Mines.* Oxford: Oxbow Books.

Torrence, R. 1986: *Production and Exchange of Stone*

Tools: Prehistoric Obsidian in the Aegean. Cambridge: Cambridge University Press.

Tyrrell, G.W. 1928: *The Geology of Arran.* Memoirs of the Geological Survey, Scotland. Edinburgh: Department of Scientific and Industrial Research / His Majesty's Stationery Office.

Vita-Finzi, C., & Higgs, E.S. 1970: Prehistoric Economy in the Mount Carmel Area of Palestine: Site Catchment Analysis. *Proceedings of the Prehistoric Society* XXXVI, 1-37.

Warren, G. 2003: *Calanais Field Project, Lewis. Chipped Stone.* Unpublished report from Edinburgh University.

Warren, G. 2006: Chipped Stone Tool Industries of the Earlier Neolithic in Eastern Scotland. *Scottish Archaeological Journal* 28(1), 27-47.

Warren, G., & Neighbour, T. 2004: Quality Quartz: Working Stone at a Bronze Age Kerbed Cairn at Olcote, near Calanais, Isle of Lewis. *Norwegian Archaeological Review* 37 (1), 83-94.

Watts, S., & Pollard, A.M. 1998: Identifying archaeological jet and jet-like artifacts using FTIR. *In* B. Pretzel (ed.): *Postprints: IRUG2 at the V&A,* 37-52. London: Victoria & Albert Museum.

Whittaker, J.C. 1994: *Flintknapping. Making and Understanding Stone Tools.* Austin: University of Texas Press.

Wickham-Jones, C.R. 1990: *Rhum. Mesolithic and Later Sites at Kinloch. Excavations 1984-86.* Society of Antiquaries of Scotland Monograph Series 7. Edinburgh: Society of Antiquaries of Scotland.

Wiessner, P. 1983: Style and Social Information in Kalahari San Projectile Points. *American Antiquity* 48 (2), 225-276.

Wiessner, P. 1984: Reconsidering the Behavioral Basis for Style: A Case Study among the Kalahari San. *Journal of Anthropological Archaeology* 3, 190-234.

Williams Thorpe, O., & Thorpe, R.S. 1984: The Distribution and Sources of Archaeological Pitchstone in Britain. *Journal of Archaeological Science* 11, 1-34.

Wilmsen, E.N. 1972: Introduction: The Study of Exchange as Social Interaction. *In* E.N. Wilmsen (ed.): *Social Exchange and Interaction,* 1-4. Museum of Anthropology, University of Michigan, Anthropological Papers 46. Ann Arbor: University of Michigan.

Wobst, H.M. 1974: Boundary Conditions for Palaeolithic Social Systems: A Simulation Approach. *American Antiquity* 39 (2), 147-178.

Wobst, H.M. 1977: Stylistic Behavior and Information Exchange. *Anthropological Papers (University of Michigan)* 61, 317-342.

Woodman, P.C., Finlay, N., & Anderson, E. 2006: *The Archaeology of a Collection: The Keiller-Knowles Collection of the National Museum of Ireland.* Bray: Wordwell Ltd. / National Museum of Ireland.

Yerkes, R.W. 2002: Hopewell Tribes: A Study of Middle Woodland Social Organization in the Ohio Valley. *In* W.A. Parkinson (ed.): *The Archaeology of Tribal Societies.* Archaeological Series 15, 227-245. Ann Arbor: International Monographs in Prehistory.

APPENDIX 1:
EXAMINED ASSEMBLAGES: BASIC DATA SORTED BY REGION AND SITE.

(for more details, see the project database, which has been lodged with the RCAHMS, Edinburgh)

Cat no	Region	Site	NGR	Total	Raw	Debitage	Prep flakes	Cores	Tools	Aphyric	Porphyritic
M043	Angus	Ardownie Farm	NO 49 34	1	0	0	0	0	1	0	1
M328	Angus	East Lochside, Kirriemuir	NO 357 548	1	0	1	0	0	0	0	1
M271	Angus	Fordhouse Barrow, House of Dun	NO 6658 6053	10	0	4	2	0	4	10	0
M035	Angus	Tannadice	NO 475 581	3	0	0	0	0	3	2	1
M027	Argyll & Bute	Acharn Farm, Morvern	NM 6975 5048	1	0	1	0	0	0	1	0
M013	Argyll & Bute	Achnacreebeag, Benderloch	NM 929 364	3	0	2	0	0	1	1	2
M182	Argyll & Bute	Ardnadam, Cowal	NS 1588 8049	1	0	0	0	0	1	1	0
M072	Argyll & Bute	Auchategan, Glendaruel, Bute	NS 002 843	90	3	63	4	14	6	61	29
M282	Argyll & Bute	Balloch Hill, Kintyre	NR 677 176	54	1	39	1	10	3	37	17
M093	Argyll & Bute	Beacharra Cairn, Kintyre	NR 6926 4334	1	0	1	0	0	0	1	0
M087	Argyll & Bute	Blackpark Plantation East, Bute	NS 093 555	240	0	219	0	6	15	21	219
M090	Argyll & Bute	Brackley, Kintyre	NR 7937 4187	4	0	3	0	1	0	4	0
M281	Argyll & Bute	Bruich an Druimen	NR 830 960	6	0	2	0	3	1	1	5
M213	Argyll & Bute	Dunagoil, Bute	NS 0846 5312	1	0	0	0	0	1	0	1
M067	Argyll & Bute	Ellary Boulder Cave	NR 7392 7649	62	0	58	0	2	2	62	0
M010	Argyll & Bute	Glecknabae Cairn (Chamber 1)	NS 007 682	1	0	0	0	1	0	0	1
M109	Argyll & Bute	Glendaruel	NS 00 84	1	0	1	0	0	0	1	0
M108	Argyll & Bute	Kilmun Arboretum	NS 164 824	1	0	1	0	0	0	1	0
M327	Argyll & Bute	Kingarth Quarry, Bute	NS 0955 5605	7	0	4	0	2	1	4	3
M212	Argyll & Bute	Kyles of Bute	NS 0443 7205	1	0	1	0	0	0	1	0
M012	Argyll & Bute	Michael's Grave	NR 994 703	1	0	0	0	1	0	1	0
M262	Argyll & Bute	Midross, Loch Lomond (Site 10/1)	NS 3590 8595	4	0	3	0	0	1	4	0
M260	Argyll & Bute	Midross, Loch Lomond (Site 5/1)	NS 3590 8595	5	0	4	0	0	1	5	0
M261	Argyll & Bute	Midross, Loch Lomond (Site 5/3)	NS 3590 8595	18	4	11	1	0	2	17	1
M086	Argyll & Bute	Shalunt, Bute	NS 051 713	1	0	1	0	0	0	1	0
M068	Argyll & Bute	St Blane's, Bute	NS 099 534	1	0	0	0	0	1	0	1
M291	Argyll & Bute	St Marnock's Chapel, Inchmarnock	NS 0236 5963	2	0	0	0	0	2	2	0
M270	Arran	Allt Lebnaskey	NS 015 294	9	2	5	0	1	1	9	0
M099	Arran	Arran	Too broad provenance	2	0	1	0	0	1	2	0
M100	Arran	Arran	Too broad provenance	1	0	1	0	0	0	0	1
M091	Arran	Arran	Too broad provenance	12	11	0	0	0	1	12	0

Cat no	Region	Site	NGR	Total	Raw	Debitage	Prep flakes	Cores	Tools	Aphyric	Porphyritic
M337	Arran	Auchareoch	NR 9952 2472	12	0	7	0	1	4	12	0
M049	Arran	Auchareoch	NR 994 245	1	0	0	0	0	1	1	0
M269	Arran	Auchrannie, Brodick	NS 008 359	48	5	40	0	2	1	4	44
M346	Arran	Balnagore	NR 901 329	2	0	0	0	1	1	0	2
M074	Arran	Blairmore	NS 026 330	1	0	0	0	0	1	1	0
M033	Arran	Carn Ban, Kilmory Water	NR 991 262	2	0	2	0	0	0	1	1
M044	Arran	Carn Ban, Kilmory Water	NR 9910 2620	5	0	1	0	2	2	0	5
M344	Arran	Corriegills	NS 042/3 346	1	0	0	0	0	1	1	0
M005	Arran	Corriegills	NS 03 34	1	0	0	0	0	1	1	0
M034	Arran	Corriegills	NS 03 34	1	0	0	0	0	1	1	0
M339	Arran	Corriegills (McKelvie Collection)	NS 03 34	5	0	3	0	0	2	5	0
M353	Arran	Corriegills (McKelvie Collection)	NS 03 34	6	0	2	0	0	4	6	0
M063	Arran	Corriegills, Kilbride	NS 03 34	13	5	5	1	0	2	13	0
M343	Arran	Craigend Farm	NS 031 208	1	0	0	0	0	1	1	0
M101	Arran	Dun Fionn	NS 046 338	2	0	1	0	0	1	2	0
M011	Arran	Dunan Beag (Blairmore)	NS 026 330	6	0	5	0	1	0	6	0
M073	Arran	Dunan Mor (Blairmore)	NS 027 331	2	0	2	0	0	0	2	0
M340	Arran	Glen Rosa	Unknown	4	0	3	0	1	0	4	0
M342	Arran	Glen Rosa	NS 004/5 375/6	6	0	2	0	2	1	6	0
M286	Arran	Kilmory	NR 960 216	6	0	6	0	0	0	0	6
M097	Arran	Kilpatrick	Too broad provenance	1	0	1	0	0	0	1	0
M190	Arran	Kilpatrick, Site 16/1	NR 908 263	195	12	127	2	7	47	139	56
M194	Arran	Kilpatrick, Site 16/11B (Fernie Bank)	NR 908 263	90	3	58	1	1	27	52	38
M195	Arran	Kilpatrick, Site 16/14 (Fernie Bank)	NR 908 263	33	2	20	0	2	9	27	6
M191	Arran	Kilpatrick, Site 16/2	NR 908 263	88	1	53	0	5	29	52	36
M192	Arran	Kilpatrick, Site 16/3	NR 908 263	32	0	21	0	3	8	6	26
M294	Arran	Kilpatrick, Site 16/7/2 (Fernie Bank)	NR 908 263	1	0	0	0	0	1	1	0
M193	Arran	Kilpatrick, Site 16/9	NR 905 261	5	0	3	0	1	1	3	2
M338	Arran	Loch Urie (path to LU from Lamlash)	NS 014 293 (?)	1	0	0	0	1	0	1	0
M042	Arran	Machrie	NR 9 3	2	0	0	0	1	1	0	2
M132	Arran	Machrie Moor	NR 9 3	2	0	2	0	0	0	1	1
M077	Arran	Machrie Moor I	NR 9120 3239	407	4	327	3	31	42	186	221
M078	Arran	Machrie Moor XI	NR 9121 3241	554	14	475	3	32	30	241	318
M306	Arran	Machrie North 24, Area 1606/6	NR 90 34	1	0	0	0	0	1	0	1
M303	Arran	Machrie North 24, Area 610	NR 90 34	3	0	3	0	0	0	2	1
M304	Arran	Machrie North 24, Area 705/5	NR 90 34	12	6	5	0	1	0	2	10
M305	Arran	Machrie North 24, Area 818	NR 90 34	1	0	1	0	0	0	1	0
M299	Arran	Machrie North 24, surface	NR 90 34	1	0	1	0	0	0	0	1
M307	Arran	Machrie North 24, Test Pit 116	NR 90 34	2	1	1	0	0	0	0	2

Cat no	Region	Site	NGR	Total	Raw	Debitage	Prep flakes	Cores	Tools	Aphyric	Porphyritic
M308	Arran	Machrie North 24, Test Pit 117	NR 90 34	16	0	13	0	2	1	0	16
M309	Arran	Machrie North 24, Test Pit 318-419	NR 90 34	1	0	1	0	0	0	0	1
M310	Arran	Machrie North 24, Test Pit 416-417	NR 90 34	8	0	5	0	3	0	5	3
M311	Arran	Machrie North 24, Test Pit 707	NR 90 34	1	0	1	0	0	0	0	1
M312	Arran	Machrie North 24, Test Pit 804	NR 90 34	1	0	1	0	0	0	1	0
M295	Arran	Machrie North 24/1	NR 9006 3460	121	12	70	0	15	24	85	36
M296	Arran	Machrie North 24/2	NR 896 343	1	0	1	0	0	0	1	0
M297	Arran	Machrie North 24/3	NR 9007 3480	20	0	14	0	3	3	11	9
M298	Arran	Machrie North 24/50	NR 899 342	23	0	12	0	6	5	20	3
M300	Arran	Machrie North 24/6B	NR 90 34	2	0	1	0	0	1	1	1
M301	Arran	Machrie North 24/7B	NR 900 346	1	0	0	0	0	1	1	0
M302	Arran	Machrie North 24/8	NR 90 34	1	0	0	0	0	1	0	1
M313	Arran	Machrie North, buried soils	NR 90 34	1	0	0	0	0	1	1	0
M139	Arran	Monamore Cairn (Meallach's Cairn)	NS 0175 2889	118	36	68	0	9	5	114	4
M347	Arran	No provenance	Unknown	3	0	3	0	0	0	3	0
M351	Arran	No provenance	Unknown	14	0	5	0	1	8	13	1
M348	Arran	No provenance	Unknown	1	0	0	0	0	1	0	1
M345	Arran	No provenance	Unknown	1	0	0	0	0	1	1	0
M350	Arran	No provenance (forestry work)	Unknown	8	0	0	0	0	8	8	0
M341	Arran	No provenance (forestry work)	Unknown	3	0	0	0	0	3	3	0
M098	Arran	Shanghai 1, Broombrae, Blackwaterfoot	NR 9029 2742	2	2	0	0	0	0	0	2
M214	Arran	Shedog Inn, Arran	Unknown	1	0	0	0	0	1	0	1
M349	Arran	Sliddery Moor Farm	NR 9295 2435	24	0	0	0	2	22	22	2
M110	Arran	Tormore	NR 897 315	89	4	76	1	2	6	1	88
M075	Arran	Tormore	Unknown	1	0	0	0	0	1	1	0
M314	Arran	Tormore 10/1	NR 896 312	14	0	7	0	3	4	2	12
M009	Arran	Tormore I	NR 903 310	2	0	2	0	0	0	1	1
M060	Arran	Torr Righ Beag, near Machrie	NR 898 315	8	0	6	0	0	2	0	8
M355	Arran	Unprovenanced	Unknown	11	0	0	0	0	11	8	3
M354	Arran	Unprovenanced	Unknown	5	0	1	0	2	2	2	3
M352	Arran	Unprovenanced	Unknown	10	0	0	0	0	10	8	2
M089	Arran	Whitehouse Farm, Lamlash	NS 024 308	1	0	0	0	0	1	1	0
M032	Arran	Whiting Bay	Too broad provenance	4	0	4	0	0	0	0	4
M359	City of Edinburgh	Edinburgh Tramscheme	NT 1652 7250	1	0	1	0	0	0	1	0
M259	Clackmannan-shire	Meadowend Farm, Kennet	NS 928 904	4	0	3	0	0	1	4	0
M080	Dumfries and Galloway	Barsalloch	NX 343 422	1	1	0	0	0	0	1	0
M199	Dumfries and Galloway	Brocklerigg	NY 114 734	3	0	3	0	0	0	3	0
M016	Dumfries and Galloway	Cairnholy I	NX 517 539	1	1	0	0	0	0	1	0

Cat no	Region	Site	NGR	Total	Raw	Debitage	Prep flakes	Cores	Tools	Aphyric	Porphyritic
M202	Dumfries and Galloway	Carronbridge	NX 869 977	2	0	2	0	0	0	2	0
M203	Dumfries and Galloway	Carzield	NX 9703 8212	2	0	2	0	0	0	2	0
M116	Dumfries and Galloway	Clayshant	NX 110 526	1	0	1	0	0	0	1	0
M174	Dumfries and Galloway	Glenluce	NX 1 5	12	0	11	0	1	0	12	0
M175	Dumfries and Galloway	Glenluce	NX 1 5	1	0	1	0	0	0	1	0
M204	Dumfries and Galloway	Glenluce	NX 1 5	2	0	2	0	0	0	2	0
M001	Dumfries and Galloway	Glenluce Sands	NX 1 5	190	3	130	6	40	11	190	0
M015	Dumfries and Galloway	Glenluce Sands	NX 1 5	1	0	0	0	0	1	0	1
M003	Dumfries and Galloway	Glenluce Sands (probably)	NX 1 5	16	1	7	0	4	4	16	0
M335	Dumfries and Galloway	Kilhern Bog Site	NX 1991 6425	21	0	15	1	0	5	21	0
M066	Dumfries and Galloway	Kirkburn, Lockerbie	NY 130 832	1	0	1	0	0	0	1	0
M114	Dumfries and Galloway	Knockdoon, Luce Sands	NX 132 551	4	0	4	0	0	0	3	1
M280	Dumfries and Galloway	Knocknab, Torrs Warren, Luce Bay	NX 12 54	280	4	239	5	18	14	280	0
M106	Dumfries and Galloway	Knockree, Wigtown	NX 43 55	1	0	1	0	0	0	1	0
M129	Dumfries and Galloway	Low and Mid Torrs, Luce	NX 1 5	1	0	1	0	0	0	1	0
M104	Dumfries and Galloway	Luce	NX 1-4 3-5	3	0	1	0	0	2	3	0
M105	Dumfries and Galloway	Luce	NX 1-4 3-5	3	0	2	0	0	1	3	0
M121	Dumfries and Galloway	Luce	NX 1-4 3-5	2	0	1	0	1	0	2	0
M002	Dumfries and Galloway	Luce Bay	NX 1 5	1	0	1	0	0	0	1	0
M094	Dumfries and Galloway	Luce Bay	NX 1-4 3-5	1	0	1	0	0	0	1	0
M122	Dumfries and Galloway	Luce Bay	NX 1-4 3-5	1	0	0	0	0	1	1	0
M123	Dumfries and Galloway	Luce Bay	NX 1-4 3-5	105	3	85	1	15	1	105	0
M127	Dumfries and Galloway	Luce Bay	NX 1-4 3-5	5	0	5	0	0	0	5	0
M181	Dumfries and Galloway	Luce Bay (Star-Bead Site)	NX 13 54	1	0	1	0	0	0	1	0
M201	Dumfries and Galloway	Luce Sands	NX 1 5	9	0	9	0	0	0	9	0
M198	Dumfries and Galloway	Luce Sands	NX 1 5	1	0	1	0	0	0	1	0
M196	Dumfries and Galloway	Luce Sands	NX 1 5	661	3	616	2	37	3	658	3
M200	Dumfries and Galloway	Luce Sands	NX 1 5	1	0	1	0	0	0	1	0
M092	Dumfries and Galloway	Luce Sands	NX 1 5	20	1	14	1	4	0	20	0
M125	Dumfries and Galloway	Luce Sands	NX 1 5	30	0	28	0	1	1	30	0

90

Cat no	Region	Site	NGR	Total	Raw	Debitage	Prep flakes	Cores	Tools	Aphyric	Porphyritic
M126	Dumfries and Galloway	Luce Sands	NX 1 5	11	0	10	0	1	0	11	0
M276	Dumfries and Galloway	Luce Sands	NX 1 5	11	0	11	0	0	0	11	0
M278	Dumfries and Galloway	Luce Sands	NX 1 5	2	0	2	0	0	0	2	0
M081	Dumfries and Galloway	Luce Sands (Pin Dune A)	NX 129 536	1	0	1	0	0	0	1	0
M273	Dumfries and Galloway	Mark Farm, Castle Kennedy	NX 11 57	2	0	2	0	0	0	2	0
M120	Dumfries and Galloway	Mid Torrs, Luce	NX 13 55	1	0	1	0	0	0	1	0
M275	Dumfries and Galloway	No provenance	Unknown	4	0	1	0	1	2	4	0
M277	Dumfries and Galloway	No provenance	Unknown	12	0	11	0	0	1	10	2
M279	Dumfries and Galloway	No provenance	Unknown	40	0	37	0	3	0	40	0
M274	Dumfries and Galloway	Portpatrick	NW 99 54	1	0	0	0	0	1	1	0
M113	Dumfries and Galloway	Star Beach, Luce Bay	NX 13 54	6	0	6	0	0	0	6	0
M107	Dumfries and Galloway	Star Site	NX 132 551	54	0	51	1	1	1	54	0
M173	Dumfries and Galloway	Torrs Warren, Dunragit	NX 14 54	2	0	3	0	0	0	3	0
M076	Dumfries and Galloway	Torrs Warren, Luce Sands	NX 128 547	179	149	3	0	22	5	177	2
M128	Dumfries and Galloway	Torrs, Luce	NX 1 5	2	0	1	0	1	0	2	0
M197	Dumfries and Galloway	Twigleas area, Eskdalemuir	NY 25 97	12	0	10	0	2	0	12	0
M088	Dumfries and Galloway	Wigtownshire	NX 2 6	1	0	1	0	0	0	0	1
M289	East Ayrshire	Laigh Newton	NS 600 370	3	0	2	0	0	1	3	0
M048	East Lothian	Crichness, Cranshaws	NT 68 66	1	0	1	0	0	0	1	0
M007	East Lothian	Hedderwick	NT 63 77	12	0	9	1	0	2	12	0
M024	East Lothian	West Links, Dirleton	NT 54 85	1	0	1	0	0	0	1	0
M331	Fife	Balfarg Riding School	NO 2850 0316	20	1	13	0	2	4	20	0
M052	Fife	Brackmont Mill	NO 436 223	4	0	2	0	0	2	4	0
M053	Fife	St Michael's Sandpit, Leuchars	NO 436 229	1	0	1	0	0	0	1	0
M284	Glasgow City	West Flank Road, Drumchapel	NS 509 712	1	0	1	0	0	0	1	0
M320	Highland (Caithness/Sutherland)	Camster Long	ND 260 442	1	0	1	0	0	0	1	0
M008	Highland (Caithness/Sutherland)	Golspie	NH 81 97	1	0	1	0	0	0	1	0
M061	Highland (Caithness/Sutherland)	Ord North, Lairg	NC 5733 0560	1	0	1	0	0	0	1	0
M017	Highland (Caithness/Sutherland)	Tulach an t'Sionnaich, Loch Calder	ND 070 619	1	0	1	0	0	0	1	0
M065	Highland (Inverness/Nairn)	Dahl House, Polloch, Glenfinnan	NM 788 683	1	0	0	0	1	0	1	0
M285	Highland (Inverness/Nairn)	Urquhart Castle	NH 5305 2860	1	0	1	0	0	0	1	0

Cat no	Region	Site	NGR	Total	Raw	Debitage	Prep flakes	Cores	Tools	Aphyric	Porphyritic
M272	Highland (Ross/Cromarty)	Achnahaird Sands	NC 0167 1314	1	1	0	0	0	0	1	0
M176	Highland (Ross/Cromarty)	Risga	NM 61 59	1	0	1	0	0	0	1	0
M136	Inverclyde	Gryfe Reservoir A	NS 2823 7147	1	0	1	0	0	0	1	0
M137	Inverclyde	Gryfe Reservoir A, Open Site 11	NS 2823 7147	2	1	1	0	0	0	2	0
M138	Inverclyde	Gryfe Reservoir A, Open Site 12	NS 2823 7147	7	1	2	0	2	2	7	0
M134	Inverclyde	Gryfe Reservoir A, Open Site 2	NS 2823 7147	5	1	3	0	1	0	5	0
M133	Inverclyde	Gryfe Reservoir A, Open Site 3	NS 2823 7147	2	0	1	0	0	1	2	0
M135	Inverclyde	Gryfe Reservoir A, Open Site 4	NS 2823 7147	1	0	1	0	0	0	1	0
M050	Midlothian	Castlelaws Fort	NT 22 65	1	0	0	0	0	1	1	0
M322	Midlothian	Elginhaugh	NT 321 673	5	0	5	0	0	0	5	0
M211	Midlothian	Newfarm, Dalkeith Northern Bypass	NT 347 688	1	0	1	0	0	0	1	0
M064	Moray	Culbin Sands	NH 9 6	4	0	1	0	0	3	4	0
M130	NE Scotland	NE Scotland	Too broad provenance	2	0	2	0	0	0	2	0
M321	North Ayrshire	Carwinning Hill	NS 2871 5286	58	0	45	1	2	10	54	4
M071	North Ayrshire	Carwinning Hill, near Dalry	NS 28 52	1	0	0	0	1	0	0	1
M079	North Ayrshire	Highthorn	NS 20 50	2	0	1	0	0	1	1	1
M180	North Ayrshire	Martin Glen Roundhouse, Largs	NS 2287 6710	1	0	1	0	0	0	1	0
M287	North Ayrshire	Meadowhead, Shewalton	NS 3389 3586	1	0	1	0	0	0	1	0
M283	North Ayrshire	Moss, near Irvine	NS 32 38	35	0	34	0	1	0	34	1
M095	North Ayrshire	Shewalton	NS 3 3	1	0	0	0	0	1	0	1
M096	North Ayrshire	Shewalton	NS 3 3	18	0	18	0	0	0	12	6
M336	Northumberland	Low Trewhett, Rothbury	NU 002 048	1	0	1	0	0	0	1	0
M085	Orkney	Barnhouse	HY 306 124	24	0	16	1	3	4	16	8
M210	Orkney	Ness of Brodgar	HY 3039 1280	1	0	0	0	0	1	1	0
M209	Orkney	Ness of Brodgar	HY 3024 1290	1	0	0	0	0	1	1	0
M318	Perth and Kinross	Aldclune, Blair Atholl	NN 894 642	1	0	1	0	0	0	0	1
M325	Perth and Kinross	Ben Lawers	NN 6639 4274	1	0	1	0	0	0	1	0
M319	Perth and Kinross	Ben Lawers	NN 6910 4288	2	0	2	0	0	0	2	0
M357	Perth and Kinross	Blaeberry, Dunning	NO 012 153	1	0	1	0	0	0	1	0
M046	Perth and Kinross	Dalnaglar	NO 15 64	1	0	0	0	0	1	1	0
M356	Perth and Kinross	Nethermuir	NO 156 411	1	0	1	0	0	0	1	0
M324	Perth and Kinross	North Mains, Strathallan (Barrow)	NN 926 162	8	0	5	0	1	2	7	1
M323	Perth and Kinross	North Mains, Strathallan (Henge)	NN 9285 1625	5	0	2	0	1	2	4	1
M014	Perth and Kinross	Pitnacree	NN 9287 5337	1	0	0	0	0	1	1	0
M329	Renfrewshire	Houston (South Mound)	NS 4009 6647	5	0	3	0	2	0	5	0
M056	Scottish Borders	Airhouse Farm, Oxton	NT 479 536	2	0	1	0	0	1	2	0
M179	Scottish Borders	Airhouse, Lauder	NT 48 53	2	0	2	0	0	0	2	0
M021	Scottish Borders	Airhouse, Oxton	NT 479 536	1	0	0	0	0	1	1	0
M131	Scottish Borders	Bedrule	NT 60 18	1	0	0	0	0	1	1	0
M185	Scottish Borders	Bedrule	NT 60 18	1	0	0	0	1	0	1	0
M184	Scottish Borders	Billerule, Bedrule	NT 60 18	2	0	2	0	0	0	2	0

Cat no	Region	Site	NGR	Total	Raw	Debitage	Prep flakes	Cores	Tools	Aphyric	Porphyritic
M151	Scottish Borders	Carfrae, Channelkirk	NT 49 54	1	0	0	0	0	1	1	0
M165	Scottish Borders	Clackmae, Melrose	NT 561 393	1	0	1	0	0	0	1	0
M140	Scottish Borders	Cowdenknowes, Earlston	NT 58 37	1	0	1	0	0	0	1	0
M057	Scottish Borders	Craigsford	NT 56 38	1	0	0	0	0	1	1	0
M047	Scottish Borders	Craigsford Mains	NT 56 38	1	0	1	0	0	0	1	0
M025	Scottish Borders	Darlingford, Earlston	NT 57 38	1	0	0	0	0	1	1	0
M158	Scottish Borders	Denholm, Cavers	NT 56 18	3	0	3	0	0	0	3	0
M018	Scottish Borders	Dryburgh (?)	NT 58 32	2	0	2	0	0	0	2	0
M019	Scottish Borders	Dryburgh Mains	NT 58 32	2	0	1	0	0	1	2	0
M059	Scottish Borders	Dryden, Ashkirk	NT 475 235	1	0	0	0	1	0	1	0
M186	Scottish Borders	Dyke, Cavers	NT 5 1	1	0	0	0	0	1	1	0
M141	Scottish Borders	Earlston	NT 57 38	1	0	0	0	0	1	1	0
M020	Scottish Borders	Earlston	NT 57 38	1	0	1	0	0	0	1	0
M006	Scottish Borders	Earlston area	NT 57 38	1	0	1	0	0	0	1	0
M170	Scottish Borders	Fairnington, Roxburgh	NT 64 28	1	0	0	0	0	1	1	0
M164	Scottish Borders	Frogden, Linton	NT 77 29	1	0	1	0	0	0	1	0
M157	Scottish Borders	Gatehousecote, Bedrule	NT 5 1	1	0	0	0	1	0	1	0
M332	Scottish Borders	Greenhill, Selkirk	NT 475 252	1	0	0	0	0	1	1	0
M143	Scottish Borders	Greenlaw	NT 71 46	4	0	4	0	0	0	4	0
M142	Scottish Borders	Halliburton, Greenlaw	NT 67 48	4	0	2	0	0	2	4	0
M153	Scottish Borders	Halliburton, Greenlaw	NT 67 48	2	0	2	0	0	0	2	0
M168	Scottish Borders	Hallrule, Bedrule	NT 59 14	2	0	1	0	0	1	2	0
M163	Scottish Borders	Hawthornside, Hobkirk	NT 56 12	2	0	2	0	0	0	2	0
M166	Scottish Borders	Hoselaw, Sprouston	NT 80 32	3	0	2	0	0	1	3	0
M146	Scottish Borders	Hume Hall	NT 71 41	1	0	1	0	0	0	1	0
M023	Scottish Borders	Kelso Area	NT 72 34	9	0	4	1	0	4	9	0
M292	Scottish Borders	Kinegar Quarry	NT 773 701	1	0	1	0	0	0	1	0
M156	Scottish Borders	Kirkton, Cavers	NT 539 138	2	0	1	0	0	1	2	0
M315	Scottish Borders	Lemington, near Reston	NT 86 63	1	0	1	0	0	0	1	0
M159	Scottish Borders	Lower Tofts, Cavers	NT 558 142	2	0	1	0	0	1	2	0
M055	Scottish Borders	Lurdenlaw, Sprouston	NT 762 320	1	0	0	0	1	0	1	0
M070	Scottish Borders	Meldon Bridge	NT 2057 4029	1	0	1	0	0	0	1	0
M189	Scottish Borders	Mertoun, Dryburgh	NT 61 32	1	0	1	0	0	0	1	0
M149	Scottish Borders	Mertoun, Dryburgh	NT 61 32	6	0	6	0	0	0	6	0
M144	Scottish Borders	Monksford Field, Mertoun, Dryburgh	NT 58 32	1	0	0	0	1	0	1	0
M026	Scottish Borders	Muirhouselaw, Maxton	NT 628 286	7	0	5	0	0	2	6	1
M169	Scottish Borders	New Graden, Linton	NT 77 29	1	0	1	0	0	0	1	0
M031	Scottish Borders	Newstead, Melrose	NT 5700 3440	3	0	1	0	0	2	2	1
M150	Scottish Borders	Orchard Field, Mertoun, Dryburgh	NT 61 32	6	0	3	0	1	2	6	0
M171	Scottish Borders	Philiphaugh, Selkirk	NT 45 28	1	0	1	0	0	0	1	0
M028	Scottish Borders	Riddleton, Maxton	NT 62 30	2	0	0	0	2	0	2	0
M040	Scottish Borders	Rink (The), Galashiels	NT 4850 3224	1	0	0	0	0	1	1	0
M333	Scottish Borders	Rink Farm	NT 485 322	4	0	3	0	0	1	4	0
M183	Scottish Borders	Riverside Field, Mertoun, Dryburgh	NT 58 32	3	0	3	0	0	0	3	0
M147	Scottish Borders	Riverside Field, Mertoun, Dryburgh	NT 58 32	2	0	2	0	0	0	2	0
M167	Scottish Borders	Roxburghshire	Too broad provenance	14	0	5	0	3	6	14	0

Cat no	Region	Site	NGR	Total	Raw	Debitage	Prep flakes	Cores	Tools	Aphyric	Porphyritic
M038	Scottish Borders	Ruberslaw	NT 553 146	1	0	1	0	0	0	1	0
M039	Scottish Borders	Ruberslaw	NT 553 146	1	0	0	0	0	1	0	1
M334	Scottish Borders	Selkirk	NT 4 2	1	0	1	0	0	0	1	0
M172	Scottish Borders	Selkirkshire	Too broad provenance	3	0	1	0	2	0	3	0
M036	Scottish Borders	Slipperfield, West Linton	NT 13 51	2	0	0	0	0	2	2	0
M022	Scottish Borders	Smedheugh, Selkirk	NT 49 27	1	0	0	0	0	1	1	0
M161	Scottish Borders	Tofts, Cavers	NT 54-5 13-4	2	0	1	0	1	0	2	0
M155	Scottish Borders	Town o'Rule, Hobkirk	NT 58 13	2	0	1	0	0	1	2	0
M160	Scottish Borders	Upper Tofts, Cavers	NT 544 137	2	0	2	0	0	0	2	0
M152	Scottish Borders	West Linton	NT 15 51	2	0	1	0	0	1	2	0
M037	Scottish Borders	West Linton	NT 15 51	1	0	0	0	0	1	1	0
M148	Scottish Borders	West Morriston, Legerwood	NT 60 40	1	0	1	0	0	0	1	0
M162	Scottish Borders	Whitriggs, Cavers	NT 56 15	13	0	7	0	2	4	13	0
M290	South Ayrshire	Gallow Hill, Girvan	NX 195 997	4	0	3	0	1	0	4	0
M288	South Ayrshire	Girvan	NX 185 985	10	0	5	0	2	3	9	1
M360	South Ayrshire	Maybole	NS 3025 0902	5	0	5	0	0	0	5	0
M051	South Lanarkshire	Berries Burn, Castle Crawford Farm	NS 951 218	1	0	0	0	1	0	1	0
M229	South Lanarkshire	Biggar Common East (Carwood Farm)	NT 005 385	73	0	64	0	5	4	73	0
M230	South Lanarkshire	Biggar Common West	NT 005 385	54	0	41	1	1	11	54	0
M264	South Lanarkshire	Boghall Farm (Biggar Gap Project)	NT 0364 3670	1	0	0	0	0	1	1	0
M217	South Lanarkshire	Brownsbank Farm, Field 4 (excav. 2000)	NT 0766 4272	61	0	54	2	3	2	61	0
M245	South Lanarkshire	Brownsbank Farm, fieldwlk 1997 (PNB)	NT 081 434	4	0	2	0	0	2	4	0
M246	South Lanarkshire	Brownsbank Farm, fieldwlk 1998 (PNB)	NT 080 433	2	0	2	0	0	0	2	0
M247	South Lanarkshire	Brownsbank Farm, fieldwlk 1999 (PNB)	NT 076 427	10	0	10	0	0	0	10	0
M248	South Lanarkshire	Brownsbank Farm, fieldwlk 1999 (PNB)	NT 074 424	8	0	7	1	0	0	8	0
M250	South Lanarkshire	Brownsbank Farm, fieldwlk 2000 (PNB)	NT 081 432	4	0	4	0	0	0	4	0
M249	South Lanarkshire	Brownsbank Farm, fieldwlk 2000 (PNB)	NT 074 427	41	0	34	1	5	1	41	0
M266	South Lanarkshire	Brownsbank Farm, fieldwlk 2002 (PNB)	NT 0717 4264	1	0	1	0	0	0	1	0
M267	South Lanarkshire	Brownsbank Farm, fieldwlk 2002 (PNB)	NT 072 426	3	0	2	0	0	1	2	1
M238	South Lanarkshire	Cala Farm (PNB)	NS 9985 4795	2	0	2	0	0	0	2	0
M237	South Lanarkshire	Carwood Farm (PNB)	NT 0295 4035	4	0	4	0	0	0	4	0
M216	South Lanarkshire	Cloburn Cairn (Cloburn Quarry)	NS 947 414	6	0	6	0	0	0	6	0
M218	South Lanarkshire	Cocklaw Farm, Elsrickle	NT 041 414	1	0	1	0	0	0	1	0
M221	South Lanarkshire	Cornhill Farm	NT 021 347	6	0	3	0	1	2	6	0
M215	South Lanarkshire	Corse Law, Carnwath (Lang Whang)	NT 018 505	67	0	53	3	4	7	66	5
M219	South Lanarkshire	Daer Valley Reservoir, Site 8	NS 9680 0715	1	0	1	0	0	0	1	0
M235	South Lanarkshire	East Gladstone Farm (PNB)	NT 0295 4228	1	0	1	0	0	0	1	0
M220	South Lanarkshire	Hangingshaw Farm	NT 003 333	4	0	4	0	0	0	4	0

Cat no	Region	Site	NGR	Total	Raw	Debitage	Prep flakes	Cores	Tools	Aphyric	Porphyritic
M263	South Lanarkshire	Heavyside Farm (Biggar Gap Project)	NT 055 375	11	0	6	0	3	2	9	2
M268	South Lanarkshire	Howburn Farm, fieldwlk 2002 (PNB)	NT 081 435	3	0	2	0	0	1	3	0
M252	South Lanarkshire	Howburn Farm, fieldwlk 2004 (PNB)	NT 082 435	15	0	13	0	0	2	12	3
M253	South Lanarkshire	Howburn Farm, fieldwlk 2004 (PNB)	NT 0790 4215	1	0	1	0	0	0	1	0
M254	South Lanarkshire	Howburn Farm, fieldwlk 2005 (PNB)	NT 080 435	20	0	15	1	2	2	19	1
M255	South Lanarkshire	Howburn Farm, fieldwlk 2005 (PNB)	NT 0739 4291	1	0	0	0	0	1	1	0
M256	South Lanarkshire	Howburn Farm, fieldwlk 2006 (PNB)	NT 081 435	9	0	8	0	0	1	9	0
M258	South Lanarkshire	Howburn Farm, fieldwlk 2007 (PNB)	NT 0677 4389	5	0	3	0	0	2	1	0
M257	South Lanarkshire	Howburn Farm, fieldwlk 2007 (PNB)	NT 081 436	1	0	0	1	0	0	1	0
M293	South Lanarkshire	Larkhall Academy	NS 760 506	4	0	2	0	1	1	4	0
M222	South Lanarkshire	Melbourne excavation, area 1	NT 086 438	101	0	87	0	8	6	97	4
M223	South Lanarkshire	Melbourne excavation, area 2	NT 086 438	3	0	3	0	0	0	3	0
M224	South Lanarkshire	Melbourne excavation, area 3	NT 086 438	1	0	1	0	0	0	1	0
M225	South Lanarkshire	Melbourne excavation, area 4	NT 086 438	1	0	1	0	0	0	1	0
M226	South Lanarkshire	Melbourne excavation, area 5	NT 086 438	3	0	1	0	2	0	3	0
M227	South Lanarkshire	Melbourne excavation, area 6	NT 086 438	4	0	4	0	0	0	4	0
M228	South Lanarkshire	Melbourne excavation, area 7	NT 086 438	1	0	1	0	0	0	1	0
M239	South Lanarkshire	Melbourne fieldwlk 1996 (PNB)	NT 085 438	2	0	1	0	0	1	2	0
M265	South Lanarkshire	Melbourne fieldwlk 2002 (PNB)	NS 9496 0889	1	0	1	0	0	0	1	0
M241	South Lanarkshire	Melbourne Wood (PNB)	NT 086 439	1	0	1	0	0	0	1	0
M236	South Lanarkshire	Muirlea Farm (PNB)	NT 0309 4124	1	0	1	0	0	0	1	0
M234	South Lanarkshire	'Probably Melbourne' (PNB)	Unknown	1	0	0	1	0	0	1	0
M243	South Lanarkshire	Scottish Woodlands Area, North (PNB)	NT 085 444	1	0	1	0	0	0	1	0
M244	South Lanarkshire	Scottish Woodlands Area, South (PNB)	NT 087 438	39	1	29	0	3	6	39	0
M111	South Lanarkshire	Stoneyburn	NS 95 19	1	0	1	0	0	0	1	0
M102	South Lanarkshire	Stoneyburn Farm, Crawford	NS 9605 1957	7	0	3	0	2	2	7	0
M251	South Lanarkshire	Toftcombs Farm, fieldwlk 2006 (PNB)	NT 057 396	3	0	1	0	2	0	3	0
M240	South Lanarkshire	Townhead Farm, Field 3 (PNB)	NT 086 450	2	0	1	0	0	1	1	1
M326	South Lanarkshire	Wellbrae	NS 9711 4010	3	0	1	0	0	2	3	0
M145	South Lanarkshire	Wester Yardhouses, Carnwath	NT 005 507	3	0	2	0	0	1	3	0
M242	South Lanarkshire	Westmill Farm (PNB)	NT 104 460	2	0	0	0	1	1	1	1
M082	South Lanarkshire	Weston	NT 03 46	1	0	1	0	0	0	1	0

Cat no	Region	Site	NGR	Total	Raw	Debitage	Prep flakes	Cores	Tools	Aphyric	Porphyritic
M231	South Lanarkshire	Weston Farm 1998, Trench 1	NT 0337 4617	25	0	24	0	1	0	25	0
M232	South Lanarkshire	Weston Farm, fieldwalking 1998	NT 026 465	51	0	41	1	5	4	50	1
M233	South Lanarkshire	Weston Farm, fieldwalking 1999	NT 034 460	29	1	20	2	3	3	27	2
M117	Southern Hebrides	Cornaig, Tiree	NL 97 47	2	0	0	0	0	2	1	1
M069	Southern Hebrides	Cul a'Bhaile, Jura	NR 549 726	1	0	1	0	0	0	1	0
M154	Southern Hebrides	Islay	NR 224 570	1	0	0	0	0	1	1	0
M177	Southern Hebrides	Islay	NR 224 570	1	1	0	0	0	0	1	0
M178	Southern Hebrides	Islay	NR 224 570	1	0	1	0	0	0	1	0
M084	Southern Hebrides	Kinloch, Rhum	NM 403 998	6	0	6	0	0	0	6	0
M062	Southern Hebrides	Lealt Bay, Jura	NR 662 902	34	0	33	0	0	1	34	0
M054	Southern Hebrides	Lealt Bay, Jura	NR 662 902	6	0	2	0	2	2	2	4
M058	Southern Hebrides	Lussa River, Jura	NR 644 873	5	0	5	0	0	0	5	0
M187	Southern Hebrides	Lussa Wood I, Jura	NR 6449 8734	67	1	62	2	2	0	67	0
M317	Southern Hebrides	Newton, Islay	NR 341 628	1	0	1	0	0	0	1	0
M188	Southern Hebrides	North Carn, Jura	NR 685 939	1	0	1	0	0	0	1	0
M316	Southern Hebrides	Sorrisdale, Area D, Isle of Coll	NM 272 638	1	0	1	0	0	0	1	0
M083	Stirling	Chapelfield, Cowie	NS 8363 8957	19	0	17	0	1	1	19	0
M004	Stirling	Claish	NN 635 065	2	0	2	0	0	0	2	0
M330	Stirling	Cowie Road	NS 816 901	2	0	2	0	0	0	2	0
M045	Stirling	West Plean	NS 81 87	1	0	0	0	0	1	1	0
M208	Uncertain	No provenance	Unknown	12	0	11	1	0	0	12	0
M205	Uncertain	No provenance	Unknown	2	1	0	0	1	0	2	0
M124	Uncertain	No provenance	Unknown	5	0	5	0	0	0	4	1
M206	Uncertain	No provenance	Unknown	2	0	2	0	0	0	2	0
M207	Uncertain	No provenance	Unknown	3	0	2	0	0	1	3	0
M119	Uncertain	No provenance	Unknown	1	0	1	0	0	0	1	0
M103	Uncertain	No provenance	Unknown	51	0	45	1	4	1	51	0
M115	Uncertain	No provenance	Unknown	3	1	2	0	0	0	3	0
M112	Uncertain	No provenance	Unknown	10	0	8	0	0	2	10	0
M041	Uncertain	No provenance	Unknown	3	0	2	0	0	1	3	0
M030	Uncertain	No provenance	Unknown	13	0	10	0	2	1	13	0
M029	Uncertain	No provenance	Unknown	1	0	1	0	0	0	1	0
M118	Uncertain	No provenance	Unknown	5	0	5	0	0	0	5	0
M358	West Lothian	Cairnpapple	NS 9872 7173	2	0	2	0	0	0	2	0
M361	Western Isles	Bharpa Langais, North Uist	NF 8359 6589	1	0	1	0	0	0	1	0

Appendix 2:
Un-Examined Assemblages: Basic Data Sorted by Region and Site

(for more details, see the project database, which has been lodged with the RCAHMS, Edinburgh)

Cat no	Region	Site	NGR	Total
S119	Aberdeenshire	Deer's Den, Kintore	NJ 784 160	4
S060	Aberdeenshire	Dyke, Culbin	NH 99 58	2
S075	Aberdeenshire	Les Murdie Road, Elgin	NJ 224 640	1
S023	Aberdeenshire	Wardend of Durris	NO 752 928	1
S009	Aberdeenshire	Warrenfield, Crathes	NO 737 966	6
S112	Angus	Balgavies Area	NO 53 51 / NO 54 51	Several
S065	Angus	Dubton Farm, Brechin	NO 583 604	2
S024	Angus	Hawkhill	NO 681 516	2
S007	Argyll & Bute	Ambrisbeg	NS 068 596	1
S011	Argyll & Bute	Ballygreggan, Drumlemble, Kintyre	NR 661 199	2
S078	Argyll & Bute	Barr River, Morvern	NM 615 562	2
S012	Argyll & Bute	East Trodigal, Drumlemble, Kintyre	NR 645 210	3
S017	Argyll & Bute	Inchmarnock	NS 022 595	1
S019	Argyll & Bute	Kerrera (Cladh A' Bhearnaig)	NM 8449 3115	Several?
S079	Argyll & Bute	Kildonan, Kintyre	NR 7806 2778	1
S077	Argyll & Bute	Loch Glashan (Medieval settlement)	NR 9168 9254	1
S008	Argyll & Bute	Macewen's Castle, Loch Fyne	NR 9158 7955	1
S026	Argyll & Bute	Monadh An Tairbh	NR 8301 9644	Unknown
S072	Argyll & Bute	Monybachach, Kintyre	NR 907 589	5
S124	Argyll & Bute	Nether Largie, Kilmartin	NR 828 976	2
S131	Argyll & Bute	Southend, Kintyre	NR 70 10	5+
S082	Argyll & Bute	The Plan, Isle of Bute	NS 092 528	Several
S128	Argyll & Bute	Torbhlaren	NR 862 943	1
S018	Argyll & Bute	Ulva Cave (A' Chrannag)	NM 4314 3843	2
S089	Arran	Arran Ring Main Water Pipeline: CTSA1	NR 9283 3309 - 9284 3314	6
S090	Arran	Arran Ring Main Water Pipeline: CTSA2	NR 9282 3272	43
S091	Arran	Arran Ring Main Water Pipeline: CTSA3	NR 9282 3270	1
S092	Arran	Arran Ring Main Water Pipeline: CTSC	NR 9395 3488 - 9438 3324	186
S093	Arran	Arran Ring Main Water Pipeline: CTSD	NS 0015 3544 - 0079 3535	1
S094	Arran	Arran Ring Main Water Pipeline: CTSE	NS 0187 3349 - 0190 3324	626
S095	Arran	Arran Ring Main Water Pipeline: CTSF	NS 0203 3017 - 0206 3100	9363
S096	Arran	Arran Ring Main Water Pipeline: CTSH	NS 0367 2193 - 0460 2237	614
S097	Arran	Arran Ring Main Water Pipeline: WBC	NR 9780 3587	1
S098	Arran	Arran Ring Main Water Pipeline: WBD	NR 9870 3600	391
S099	Arran	Arran Ring Main Water Pipeline: WBF	NS 0079 3532 - 0101 3533	17
S100	Arran	Arran Ring Main Water Pipeline: WBG	NS 1401 3539 - 0156 3533	5
S101	Arran	Arran Ring Main Water Pipeline: WBH	NS 0206 3525	1947
S102	Arran	Arran Ring Main Water Pipeline: WBI1	NS 0193 3319	94
S103	Arran	Arran Ring Main Water Pipeline: WBI2	NS 0208 3292	9
S104	Arran	Arran Ring Main Water Pipeline: WBJ	NS 0206 3100 - 0204 3130	1
S105	Arran	Arran Ring Main Water Pipeline: WBS2	NR 9842 2159 - 9880 2148	1
S030	Arran	Auchareoch	NR 9952 2472	418
S031	Arran	Auchencairn 17	NS 0130 2920	Much
S032	Arran	Auchencairn 18	NS 0115 2893	Much
S033	Arran	Auchencairn 19	NS 0095 2860	1
S034	Arran	Auchencairn 21	NS 0349 2675	64

Cat no	Region	Site	NGR	Total
S054	Arran	Bridge Farm, Machrie Moor	NR 926 321	Much
S113	Arran	Crochandoon, near Tormore	NR 897 315	Several
S001	Arran	East Bennan, Bennan Head	NR 9935 2075	1
S066	Arran	Glen Rosa	NR 9865 3837	Several
S047	Arran	Kildonan	NS 031 208	Several
S035	Arran	Kilmory Water 6	NR 985 232	Several
S022	Arran	Kilpatrick	Unknown	2
S021	Arran	Kilpatrick	Unknown	14
S020	Arran	Kilpatrick	Unknown	25
S048	Arran	Machrie	NR 901 329	At least 77
S036	Arran	Machrie Moor 1	NR 9070 3247	19
S056	Arran	Monamore Cairn (Meallach's Cairn)	NS 0175 2889	Several
S067	Arran	Old Deer Park, Brodick	NS 006 374	Several
S049	Arran	Porta Leacach, Dippin	NS 041 214	Much
S051	Arran	Sliddery	NR 937 236	Several
S050	Arran	Sliddery	NR 935 242	Much
S052	Arran	South Glen Rosa	NS 002 369	Much
S037	Arran	The Ross 22	NR 981 290	Several
S062	Arran	Tormore	NR 9 3	1
S061	Arran	Tormore	NR 9 3	2
S063	Arran	Tormore Farm	NR 8945 3242	Several
S122	City of Edinburgh	Cramond	NT 189 767	5
S115	City of Edinburgh	Ratho, near Edinburgh	NT 1311 7100	1
S085	Co. Antrim, Northern Ireland	Ballygalley	D 371 077, D374 075	500+
S074	Co. Antrim, Northern Ireland	Craigmacagan, Rathlin Island	D 154 500	2
S088	Co. Antrim, Northern Ireland	Cushendall	Unknown	1
S086	Co. Antrim, Northern Ireland	Knockans South, Rathlin Island	D 130 515	Several
S084	Co. Antrim, Northern Ireland	Lyles Hill, Toberagnee	J 248 929	2
S083	Co. Antrim, Northern Ireland	Nappan	D 289 236	1
S087	Co. Dublin, Ireland	Lambay Island	Unknown	2
S121	Cumbria	Blackfriar Street, Carlisle	NY 39 56	2
S107	Dumfries and Galloway	Beckton Farm	NY 1305 8245	8
S003	Dumfries and Galloway	Buittle Castle	NX 8191 6162	1
S013	Dumfries and Galloway	Cairnderry	NX 315 799	2
S073	Dumfries and Galloway	Fox Plantation (SNIP)	NX 1173 5733	4
S006	Dumfries and Galloway	Holywood North	NX 9502 8012	1
S005	Dumfries and Galloway	Pict's Knowe	NX 9538 7213	20
S058	Dumfries and Galloway	Wigtownshire	NX 2 6	Much
S057	Dumfries and Galloway	Wigtownshire	NX 2 6	1
S118	Fife	Cowdenbeath	NT 1634 9325	1
S071	Fife	Devil's Burdens, West Lomond Hill	NO 193 062	2
S127	Fife	Scotstarvit Covert, Cupar	NO 3609 1093	4
S126	Fife	Tarvit Farm, Cupar	NO 388 135	1
S015	Highland (Caithness/Sutherland)	Cuthill Links, Dornoch	NH 743 873	1
S016	Highland (Inverness/Nairn)	Castlehill, Cauldfield Road, Inverness	NH 697 440	1
S076	Highland (Inverness/Nairn)	Kinbeachie, Black Isle	NH 626 625	4
S027	Highland (Inverness/Nairn)	Upper Cullernie	NH 7287 4774	Unknown
S080	Isle of Man	Ballachrink, Jurby	SC 3932 0018	1
S114	North Ayrshire	Leven, Loudon Hill,	NS 606 373	2
S053	North Ayrshire	Station Brae, Dreghorn	NS 3518 3830 - 3539 3844	Unknown
S064	Northumberland	Bowden Doors, near Belford	NU 077371	1
S029	Orkney	Ness of Brodgar	HY 30 13	1
S125	Perth and Kinross	Ben Lawers (East Tombreck)	NN 6554 3740	At least 3
S002	Renfrewshire	Bishopton, Whitemoss	NS 4182 7208	2
S129	Renfrewshire	Lurg Moor, Site A	NS 295 736	At least 5
S038	Scottish Borders	Eddleston	Various (see comments)	At least 5
S116	Scottish Borders	Foulden Moorpark, Berwickshire	NT 922 575	4

Cat no	Region	Site	NGR	Total
S039	Scottish Borders	Parkgatestone Hill	NT 086 355	Several
S 132	Scottish Borders	Rutherford	NT 64 30	1
S028	Scottish Borders	Sheriff Muir	NT 200 400	Several?
S059	Scottish Borders	Yarrow	NT 35 28	1
S055	South Ayrshire	Ailsa View (Longhill)	NS 319 186	Unknown
S130	South Ayrshire	Ballantrae	NX 083 817	4
S004	South Ayrshire	Crossraguel Abbey	NS 273 085	2
S120	South Ayrshire	Monktonhead Farm	NS 354 282	2
S068	South Lanarkshire	(Easter) Sills	NS 930 423	Unknown
S069	South Lanarkshire	Annieston	NS 992 375	Unknown
S042	South Lanarkshire	Bagmoors Farm	NS 904 442	Several
S108	South Lanarkshire	Blackshouse Burn (Swaites Hill Field D)	NS 95 41	1
S110	South Lanarkshire	Brown Hill	NS 678 338	Unknown
S045	South Lanarkshire	Castledykes	NS 928 442	1
S046	South Lanarkshire	Cloburn Quarry	NS 947 415	Several
S106	South Lanarkshire	Cloburn Quarry, Cairngryffe Hill	NS 947 414	2
S044	South Lanarkshire	Dykefoot Farm	NT 0262 5243	1
S041	South Lanarkshire	Green's Farm	NT 020 470	Several
S043	South Lanarkshire	Hillhead Farm	NS 982 405	Several
S123	South Lanarkshire	Newton Farm, Cambuslang	Unknown	Several
S070	South Lanarkshire	Powbrone Burn	NS 680 339	Unknown
S040	South Lanarkshire	Sheriffflats (Thankerton)	NS 974 378	1
S014	Southern Hebrides	Camas Daraich, Point of Sleat, Skye	NG 567 000	1
S109	Southern Hebrides	Craigfad, Islay	NR 232 557	1
S081	Southern Hebrides	Eigg 1, Eigg	NM 4 8	2
S010	Southern Hebrides	Glengarrisdale, Jura	NR 643 968	7
S111	Stirling	Dumyat Hill	NS 815 979	Several
S025	Stirling	Duntreath Standing Stones	NS 533 807	2
S117	Western Isles	Calanais, Isle of Lewis	NB 2140 3280	2

COLOUR PLATES

Plate 1. A tabular piece of aphyric and spherulitic pitchstone from the so-called Magmatic Rolls Quarry, immediately south of Brodick (Tomkeieff 1961, 10). This variety is a sub-type of the pitchstone making up the Great Sill at Dun Fionn, near Corriegills.

Plate 2. A large flake of lightly porphyritic pitchstone from Judd's dyke Tormore IV (Judd 1893, 557) on Arran's west-coast.

Plate 3. An irregular block of coarsely porphyritic pitchstone from the outcrop at Glen Shurig Farm, immediately west of Brodick.

Plate 6. A small piece of raw jet from Whitby in Yorkshire.

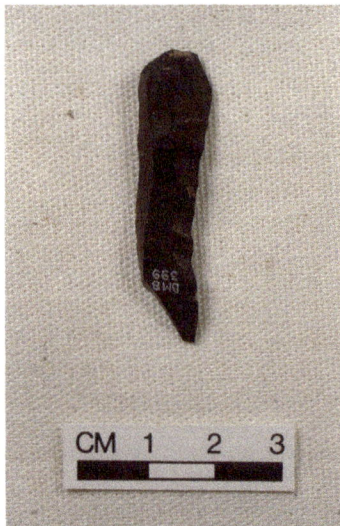

Plate 4. A blade of fine-grained black chert from Dryburgh Mains, Scottish Borders.

Plate 7. A lump of glassy slag from Loanhead of Daviot, Aberdeenshire.

Plate 5. A transverse arrowhead (Clark's Type G; Clark 1934b, 35) in black flint from Glenluce Sands, Dumfries & Galloway.

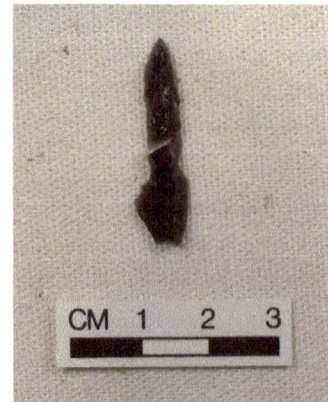

Plate 8. Refitting microlith and microburin in smoky quartz from Corse Law near Biggar, South Lanarkshire.

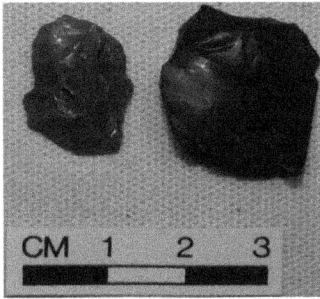

Plate 9. A light-green, burnt and 'micro-crazed' flake from Biggar Common, South Lanarkshire, (left) and a stray, unaltered, black pitchstone flake from Arran (right).

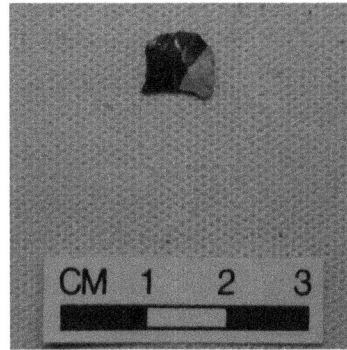

Plate 12. A partially burnt (light-brown/white) chip from Cloburn.

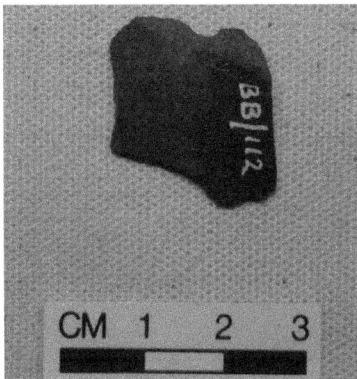

Plate 10. A light-brown, disintegrating piece from Brownsbank Farm, near Biggar in South Lanarkshire.

Plate 13. Burnt tabular pitchstone from Torrs Warren, Dumfries & Galloway.

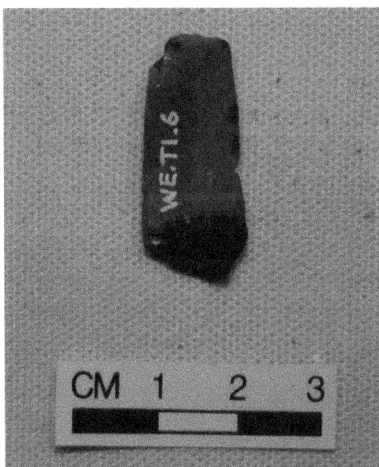

Plate 11. A partially burnt (light-brown/black) blade from Weston.

Plate 14-15. A burnt tabular piece of obsidian from the Capulin Quarry, New Mexico (Steffen 2005, 74; illustrations courtecy of Anastasia Steffen, University of New Mexico).

Plate 16. Burnt pitchstone artefacts from Barnhouse, Orkney. The blade to the left shows clear fire-enhanced banding.

Plate 17. A weathered grey pitchstone flake from Daer Reservoir. It is possible to see the original black colour where the edges have been nicked.

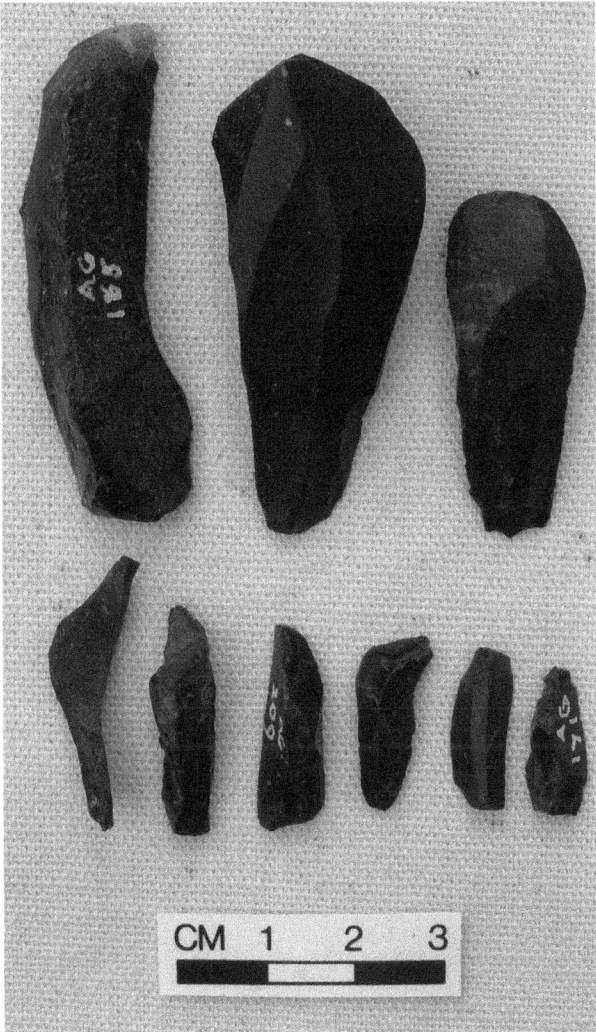

Plate 18. A selection of blades and microblades from Auchategan, Argyll & Bute.

Plate 19. A selection of typical platform-cores from Auchategan, Argyll & Bute.

Plate 20-21. A Levallois-like core from Machrie Moor, Arran. Seen from opposed faces.

Plate 22-23. Two small discoidal cores of Glen Luce Type, both from Glenluce Sands. Seen from opposed faces.

*Plate 24. Two leaf-shaped arrowheads, one (left) from the
settlement site of Auchareoch, Arran, and the other (right) from
the chambered tomb at Blairmore, Arran.*

Plate 25. Three chisel-shaped arrowheads (bottom) from Machrie Moor, Arran, and one (top) from Glenluce Sands, Dumfries & Galloway.

Plate 26. Two small end-scrapers, one (left) from Glenluce Sands, Dumfries & Galloway, and the other (right) from Allt Lebnaskey in the Monamore area, Arran.

Plate 28. A heavily curved blade from Melbourne, near Biggar in South Lanarkshire.

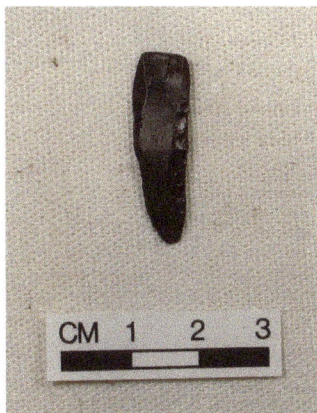

Plate 27. A microblade piercer, possibly from Culbin Sands, Aberdeenshire.

Plates 29-30. The two opposed faces of a small Levallois-like core. From Barnhouse on Orkney.

www.ingramcontent.com/pod-product-compliance
Lightning Source LLC
Chambersburg PA
CBHW061006030426

42334CB00033B/3377

www.ingramcontent.com/pod-product-compliance
Lightning Source LLC
Chambersburg PA
CBHW061006030426

42334CB00033B/3378